MORPHOMETRICS

Edited by **Christina Wahl**

Morphometrics
Edited by Christina Wahl

Published by InTech

Janeza Trdine 9, 51000 Rijeka, Croatia

Publishing Process Manager Iva Simcic

Technical Editor Teodora Smiljanic

Cover Designer InTech Design Team

Additional hard copies can be obtained from orders@intechopen.com

Morphometrics, Edited by Christina Wahl
p. cm.
ISBN 978-953-51-0172-7

For EU product safety concerns:
IN TECH d.o.o., Prolaz Marije Krucifikse Kozulić 3, 51000 Rijeka, Croatia,
info@intechopen.com or visit our website at intechopen.com.

Meet the editor

Christina Wahl earned her doctorate in Physiology at Cornell University in 1990. She studies craniofacial development, in particular differentiation of the eye and periocular structures. Other research interests include the visual ecology of larval and juvenile fishes, the effects of light regimes on sensory system development, and variations in the normal morphology of vertebrate ovaries. She teaches courses in anatomy and physiology, developmental biology, and vertebrate zoology at Wells College in Aurora, New York. She also maintains a courtesy appointment as Associate Professor in Biomedical Sciences at Cornell University.

Contents

Preface

It is human nature to measure things, and this holds true for science as well as everyday life. The five papers in this book demonstrate the usefulness of a morphometric approach to a variety of subjects in natural history, including systematics, phenotypic plasticity in response to environmental variation, and ontogenetic adaptation.

As our understanding of genetic control mechanisms and epigenetics has matured over the last several decades, it has become clear that morphometric assessment continues to be important to our overall understanding of natural variability in growth and form. The tremendous growth of our knowledge base during the last century has necessitated that we find new ways to measure and track greater detail as well as greater numbers of parameters among populations and individuals.

Computing has advanced morphometrics significantly, allowing ever larger data set analyses with ever increasing ease as technology has steadily improved. In his review of bumble bee morphometrics, Robin Owen explains the differences between "classical morphometrics", in which multiple characteristics are measured and used descriptively but not analyzed statistically, "traditional morphometrics", utilizing multivariate statistical analyses of data sets derived from point-to-point measurements such as wing length, and "geometric morphometrics", that allows comparisons of the shapes of whole organisms, or parts of organisms.

From their study of mosquito biodiversity, Jean-Pierre Dujardin, P. Thongsripong, and A. Henry contend that "when grouping many traits morphological variation becomes too complex to validly reflect a single factor like species richness." They point out that organismal morphology is the result of environmental and historical, as well as biological causes, and they propose a way to decompose morphology into elementary units and test that variation against species richness.

Tackling a completely different problem, Nikhil Whitaker documents parameters of clutch and egg sizes of the endangered Travancore tortoise, *Indotestudo travancorica,* in hopes that this information will help those who intend to breed these animals in captivity.

Paraskevi Karachle and K. Stergiou explore an allometric approach to morphometry in fishes. They correlate their findings to habitat and feeding behavior, and show that for some fishes, environment plays a significant role in determining different body metrics during post-hatching growth.

The questions being addressed using morphometric techniques have begun to change. Many of today's investigations are concerned with interactions between species and their environments. Owen reminds us that assessment of fluctuating asymmetry is a valuable way to evaluate environmental stress. In my chapter on vertebrate development, I present some examples of phenotypic variation arising from environmental stress during critical periods. I point out that while some instances of adaptive plasticity during development are adaptive…for instance, the "thrifty phenotype" of offspring born to malnourished mothers….under some circumstances, this plasticity becomes maladaptive.

We are reaping the benefits of combining what we learn through morphometry with what we learn from molecular biology, however, much remains to be done. For instance, the remarkable regenerative capacity among urodele amphibians has been documented since the 1800's, but only now are the genetic mechanisms being worked out. In the not-too-distant future, we will be able to pinpoint the biochemical pathways being activated (or deactivated) in response to specific environmental cues, and we'll be able to quantify what happens through advanced morphometric techniques. Hopefully, this information will ultimately be useful towards improved stewardship of our natural resources.

Assoc. Prof. Christina Wahl
Wells College, Aurora, New York,
USA

1

Applications of Morphometrics to the Hymenoptera, Particularly Bumble Bees (*Bombus*, Apidae)

Robin E. Owen
Department of Chemical & Biological Sciences, Mount Royal University, Calgary, Alberta, Canada

1. Introduction

I will first briefly review the types and range of morphometric studies of the Hymenoptera, and will discuss the characters used. Wing venation characters are very commonly employed and I will briefly discuss wing development and functional aspects of hymenopteran wings in this context. The chapter will be partly a selected review of work in this area by myself and others but will also include some original work of my own not previously published.

The Hymenoptera are an extremely diverse order of insects containing 144,695 described, extant species (Huber, 2009), fewer than the Coleoptera (beetles) and Lepidoptera (moths and butterflies), however if undescribed species are included then the Hymenoptera may be the most specious of all insect orders and there could be as many as a million species (Sharkey, 2007). There are two main groups of the Hymenoptera; the more primitive Symphyta (sawflies, horntails) and the Apocrita, which contain 93% of the species (Huber, 2009.) The Apocrita is subdivided into the Parasitica (parasitoids) and the Aculeata, the stinging Hymenoptera which includes the familiar ants, bees and wasps. There are many evolutionary and taxonomic questions concerning the Hymenoptera which can be answered using applications of morphometrics.

Morphometrics can be broadly defined as the quantitative study of the size and shapes of organisms. Often only parts (e.g. limbs) or organs of an organism are measured, and more general conclusions are drawn about evolutionary relationships, for example, from these measurements. What is now called *traditional morphometrics* or multivariate morphometics, is the application of multivariate statistical techniques (e.g. discriminate function analysis) to morphological data sets (Adams et al., 2004). One problem, in addition to others, with using standard multivariate methods for the analysis of shape is that linear distances are usually highly correlated and so much effort was expended correcting for size (Adams et al., 2004). The "Geometric Morphometric Revolution" overcame these problems by developing methods which allowed the shape of parts, or of the whole organism to be analysed (Rohlf & Marcus, 1993; Adams et al,. 2004). This is *geometric morphometrics.*

Morphological measurements of insects, including Hymenoptera and especially the eusocial species, have had a long history of use (e.g. Huxley 1972) and have often been termed *morphometrics*. This is not true multivariate morphometrics as currently defined above and often only involves plots of two variables, such as head width and antennal scape length to describe allometric growth and caste differences in ants (Huxley, 1972; Wilson, 1971), although a combination of univariate and multivariate statistics has sometimes been employed to determine caste differences (e.g. Gelin et al., 2008). In other studies, such as those on bees, multiple characters will be measured and used descriptively but multivariate statistical analysis is not employed. I will refer to this approach as *classical morphometrics*.

2. Morphometric studies of hymenoptera

2.1 Wings and wing venation characters

Classical morphometric studies have primarily used various mouthpart measurements in addition to a measure of overall size usually radial cell length or total length of the wing (Medler, 1962; Pekkarinen, 1979; Harder, 1985), however wing measurements alone have been used in the majority of traditional and geometric morphometric studies. In holometabolous insects the longitudinal veins develop first, followed by the crossveins. Wing veins contain trachea, blood lacunae and nervous tissue, and are sensitive to developmental disturbances, as shown by studies of *Drosophila* (Marcus, 2001). The primary function of the wing veins is to provide structural support and the pattern of venation is a crucial determinant of flight mechanics. During flight insects constantly adjust wing camber for optimal air flow, and this adjustment results from the flexural stiffness of the wing, which in turn depends on the position of the crossveins (Marcus, 2001). The pattern of venation can be quantified by measuring the coordinates of the junctions (which I will call *points*) of the longitudinal and the crossveins, which presumably reflect phylogenetic and developmental information. Wing morphometrics has been successfully used in taxonomic studies of Hymenoptera to differentiate between closely related taxa, and has also shown significant differences in wing shape, size and mechanical properties between species (Aytekin et al., 2007), however there are only a relatively few studies using wing morphometrics to estimate fluctuating asymmetry.

Essentially the same set or a slightly reduced set of coordinates have been employed in most studies of Hymenoptera. Forewings have been used in all studies but some have also used data from the hindwings (Aytekin et al., 2003, 2007; Klingenberg et al. 2001). Representative examples of hymenopteran forewings are shown in figure 1 (bumble bee, *Bombus*), figure 2 (solitary wasp, *Sphex*), figure 3 (social wasp *Dolicovespula*) and figure 4 (parasitoid wasp, Braconidae) with the points used for measurement. The wing venation in figure 1 is essential homologous among bees (Table 1) and a maximum of 20 points in any particular study have been used on the forewing (Table 1) and six on the hindwing (Aytekin et al., 2003, 2007; Klingenberg et al., 2001). There is a slight difference in the venation between bumble bees and honeybees which means that point 23 is not homologous. How the measurements are then analysed depends on the approach, e.g. traditional or geometric morphometrics, etc. The wing venation in figure 2 is homologous among some of the aculeate wasps (Table 1).

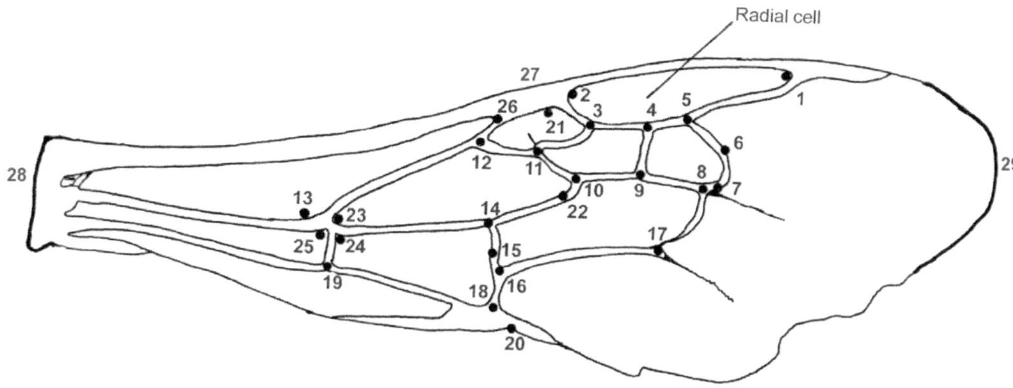

Fig. 1. Right forewing of a *Bombus rufocinctus* queen. This shows the total of 29 points which have been used in various combinations for multivariate morphometric studies of bees (see Table 1). The wing venation and the points are homologous among taxa of bees. The numbering of the first 20 points follows Aytekin et al. (2007). The length (distance 1-2) of the radial (= marginal cell) cell is also indicated as this has been used as one measure of size in some studies. The distance from the tegula (point 28) to either the distal end of the radial cell 1) or to the wingtip (29) have also been used a measures of bee size.

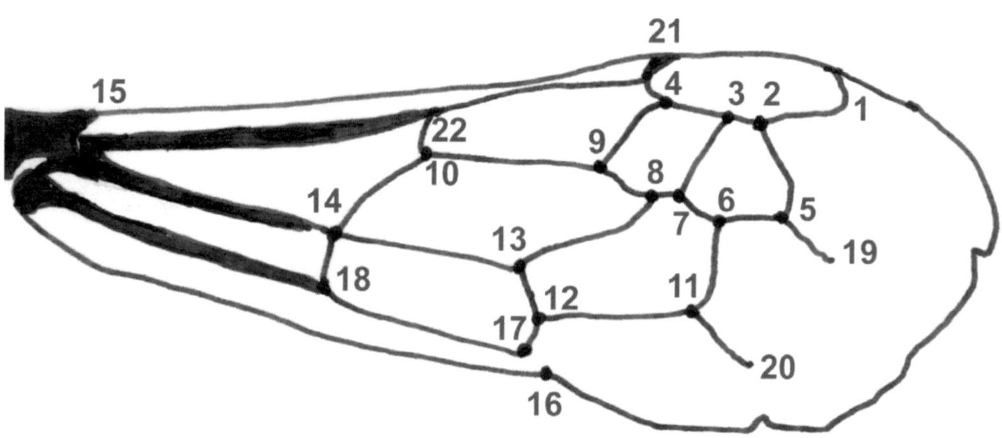

Fig. 2. Forewing of *Sphex maxillosus* redrawn from the photograph (Figure 1) of Tüzün (2009). This shows the total of 22 points which have been used in various combinations for multivariate morphometric studies of wasps (see Table 1). The wing venation and the points are homologous among taxa of aculeate wasps. The numbering of the first 20 points follows Tüzün (2009).

Family, Tribe or subfamily	Genus	Number of points	Points used (forewing)	Caste/ Sex[1]	Type of study[2]	Reference
Apidae, Bombini	*Bombus*	20	Fig. 1 : 1-20	M	G, C	Aytekin et al. 2007
Apidae, Bombini	*Bombus*	20	Fig. 1: distances (28-29),(20-27), (1-27),(1-5), (3-16),(10-13),(9-10),(3-12)	Q ,W	T, C	Aytekin et al. 2003
Apidae, Bombini	*Bombus*	19	Fig. 1: 1,2,3,4,5,8,9,10,11,12,13,14,16,18,19,21, 23,24,25	Q	T, NT	Plowright & Stephen, 1973
Apidae, Bombini	*Bombus*	19	Fig. 1: 1,2,3,4,5,8,9,10,11,12,13,14,16,18,19,21, 23,24,25	Q	T, C	Plowright & Pallett, 1978
Apidae, Bombini	*Bombus*	19	Fig. 1: 1,2,3,4,5,8,9,10,11,12,13,14,16,18,19,21, 23,24,25	Q	T, C	Plowright & Stephen, 1980
Apidae, Bombini	*Bombus*	19	Fig. 1: 1,2,3,4,5,7,8,9,10,11,12,13,14,16,17,18, 19,23,26	Q, W	T, C	Kozmus et al., 2011
Apidae, Bombini	*Bombus*	14	Fig. 1: 1,3,4,5,8,9,10,11,12,16,17,18,19,24	Q	T, C	Owen et al., 2010
Apidae, Bombini	*Bombus*	13	Fig. 1: 3,4,5,7,8,9,10,11,12,14,17,18,19,24	W	G, FA	Klingenberg et al., 2001
Apidae, Apini	*Apis*	19	Fig. 1: 1,2,3,4,5,7,8,9,10,11,12,13,14,16,17,18, 19,23,26	W	G, FA	Smith et al., 2007
Apidae, Apini	*Apis*	19	Fig. 1: 1,2,3,4,5,7,8,9,10,11,12,13,14,16,17,18, 19,23,26	W	G, ABIS, C	Francoy et al., 2009
Apidae, Euglossini	*Euglossa, Eulaema*		Fig. 1: distances M1 (1-17), M2 (1-12), M3 (12-18), M4 (17-18)	M	FA	Silva et al., 2009
Sphecidae, Sphedini	*Sphex*	20 - 24	Fig. 2: 1-20	?	G, C	Tüzün, 2009
Sphecidae, Larini	*Tachysphex*	15	Fig. 2: 1,2,3,4,6,7,8,9,10,13,14,17,18,21,22	M, F	G, C	Pretorius, 2005
Vespidae, Polistini	*Polistes*		Fig. 2: distances (1-4),(7-11), (10-12), (13-14)	Q, M	T, V	Eickwort, 1969
Vespidae, Vespinae	*Dolichovespula*	17	Fig. 3: 1-17	M	T,C	Tofilski, 2004
Braconidae, Agathidinae	*Bassus*	15	Fig. 4: 1-15	F	G, C	Baylac, et al. 2003

[1] *Q = queen, W = worker (female), F = female, M = male, ? = sex not specified.*
[2] *G = geometric morphpmetics, T = traditional morphpmetics, ABIS = automated bee identification system, FA = fluctuating asymmetry, C = classification/taxonomy, NT = numerical taxonomy, V = quantitative variation.*

Table 1. A representative selection of multivariate morphometric studies of the Hymenoptera. Either the distances between points, the distance of each point from an origin, the Cartesian coordinates of the points, or the angles between certain points are used as data. See the text for details of each study.

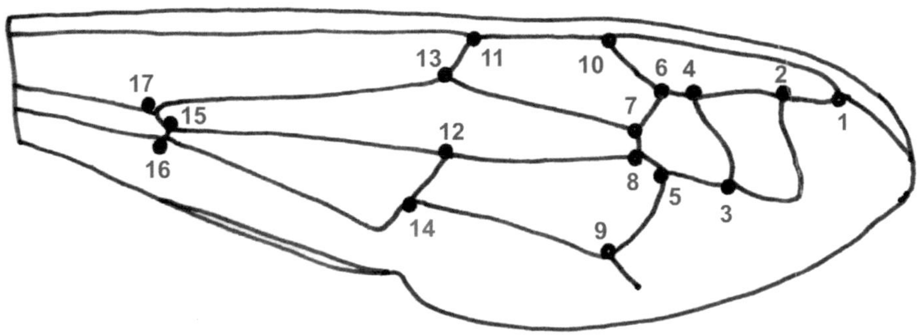

Fig. 3. Right forewing of a *Dolicovespula sylvestrimale* redrawn from figure 4 of Tofilski (2004).

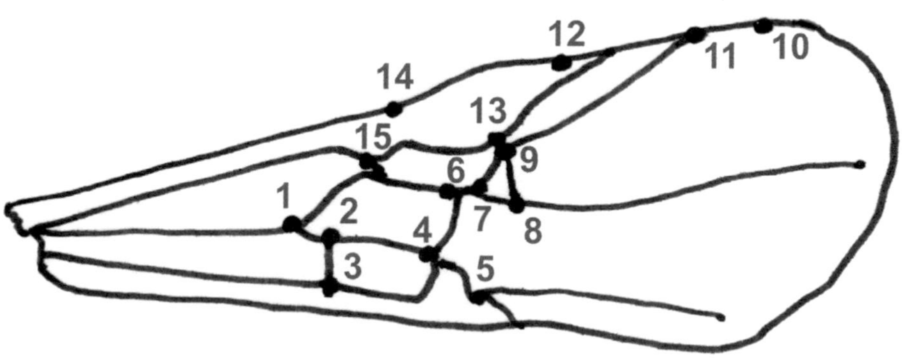

Fig. 4. Right forewing of a *Bassus tumidulus* female redrawn from figure 2 of Baylac et al. (2003).

As can be seen from the figures and Table 1, the wing venation and the wing points are not homologous among even all the Apocrita represented here, so morphometric comparisons have to be done on relatively closely related species. Wing vein characters are, however, used in cladistic analysis (Alexander, 1991; Sharkey & Roy, 2002; Shih et al., 2010) although these are generally not quantitative but instead presence/absence of veins, etc. As Sharkey & Roy (2002) point out reduction and loss characters are difficult to code and are subject to homoplasy.

2.2 Classical morphometrics

Medler (1962) measured the lengths of the radial cell of the forewing, the glossa, the prementum and the first segment of the labial palpus in 35 species of bumble bees (*Bombus* spp.). He then calculated correlation coefficients between each of these characters and calculated a wing index and a labial index (queen/worker x 100). Medler (1962) found that these indices did vary among the recognized subgenera of *Bombus*. Univariate measures of various characters of bumble bees and correlations between characters have been reported in other studies (Pekkarinen, 1979; Harder 1982, 1985; Owen, 1988). Pekkarinen (1979) measured radial cell length and calculated mouthpart indices for 13 species of bumble bees in Denmark and Fennoscandia. He found that many closely related species, subspecies or populations could be distinguished from one another on the basis of mouthpart indices (mouthpart length/radial cell length). He also found allometric variation of wing length and some mouthpart indices with body size (Pekkarinen, 1979).

Morphometric variation in relation to foraging and resource partitioning has been extensively studied in bumble bees, but these have been limited to univariate measures or indices of characters important for the foraging behaviour of worker bees. It is well established that glossa (tongue) length is a major determinant of flower choice as there is a positive correlation between glossa length and corolla length of flowers visited (Pekkarinen, 1979; Harder, 1985; Prŷs-Jones & Corbet, 1987). However Harder (1985) found that besides glossa length other factors, such as body size, wing length flower species richness and plant abundance, also influence flower choice.

Similar morphometic studies have also been done with other bees, for example stingless bees, the Meliponinae (Danaraddi & Viraktamath, 2009), and univariate measures of size variation, usually in relationship to sex ratios and sex allocation, is well known in leafcutter bees, particularly *Megachile* (e.g. Rothschild, 1979; O'Neill et al. 2010). I am not including here studies of quantitative genetic variation and heritability as these will be discussed later.

2.3 Traditional morphometrics

Discriminant function analysis, introduced by Fisher (1936), has been widely used in traditional morphometrics. Discriminant analysis is used to classify individuals into groups, i.e. to define group boundaries (Sneath & Sokal, 1973; Hintze 1996). It derives linear functions of the measurements which best discriminate populations (Fisher, 1936). These maximize discrimination between groups, the goal being to be as certain as possible that individuals are assigned to the "correct" group according to a qualitative predictor variable. Mathematically the technique is similar to multiple regression analysis, the difference being that in discriminant analysis the dependent variable is discrete instead of continuous (Hintze 1996). The predictor variable in taxonomic studies is species name, and the null hypothesis is that the original classification of the species is correct. Since discriminant analysis derives equations that maximize distinction between groups it is an inherently conservative technique as this will correspondingly minimize the likelihood of making a Type I error. Where real differences do exist the technique does correctly discriminate between species. Canonical variates analysis is very similar to discriminant function analysis except that the discriminate scores, D, are plotted in a system of orthogonal axes, which are the canonical variates (Sneath & Sokal, 1973). Discriminant functions are relatively insensitive to overall size differences, but an individual of the same shape but of much different size may be classified incorrectly (Sneath & Sokal, 1973).

Traditional morphometric approaches have been applied to problems of taxonomy, classification and geographic variation in the honeybee *Apis mellifera* and the other three commonly defined species; *A. florea*, *A. cerana*, and *A. dorsata* (Ruttner, 1986). A combination of discriminant function analysis, principal component analysis and cluster analysis allows the 23 geographic races of *A. mellifera* to be distinguished (Ruttner, 1986). Forty morphological characters were used for this analysis including angles of wing venation (Ruttner, 1986). I shall not attempt to review the large literature on honeybee morphological variation, instead I will concentrate mainly on some examples from bumble bees.

An early application of traditional morphometrics and numerical taxonomy to bumble bees was the study of Plowright & Stephen (1973) on the evolutionary relationship of *Bombus* and their social parasites, *Psithyrus*. They measured the coordinates of 19 points using point 19 as the origin and the line from 19-4 as the horizontal axis (Fig. 1, Table 1). The measurements were standardized by dividing them by the length 19-4 to give variables independent of size (Plowright & Stephen, 1973). The generalized Mahalanobis distance, D^2, was calculated for each species pair (Sneath & Sokal, 1973) and each distance was subtracted from the largest distance to give a measure of similarity (Plowright & Stephen, 1973). Plowright & Stephen (1973) then used weighted-pair-group cluster analysis (Sneath & Sokal, 1973) to produce a phenogram. The 13 species of *Psithyrus* were clearly separated from the 60 *Bombus* species. They also used multiple discriminant analysis (canonical variates analysis) to visualize the groupings (Hintze, 1996). Again *Psithyrus* was clearly separated from the *Bombus* subgenera on the plot of the first two canonical variates (Plowright & Stephen, 1973).

Traditional morphometrics has also been successful for lower level species discrimination. As will be described later, there are numerous taxonomic problems in the genus *Bombus* concerning the exact relationship of closely related species. Plowright & Pallett (1978) applied the same measurement techniques as used by Plowright & Stephen (1973) and discriminant analysis to re-investigate the taxonomic status of *B. sandersoni* Fkln. They measured previously identified museum specimens, and found a non-overlapping separation between *B. sandersoni*, and *B. frigidus* F. Sm., and *B. vagans* F. Sm. Therefore Plowright & Pallett (1978) suggested retaining *sandersoni* as the valid name for the species. However they did also point out that their results did not preclude this taxon from being a clinal variant of *frigidus*. Similarly Plowright & Stephen (1980) re-examined the taxonomic status of *Bombus franklini* (Frison) and multivariate analysis gave a clear separation of *franklini* from other species within the subgenus.

Tofilski (2004) was able to correctly classify all 22 individuals of the two wasp species *Dolicovespula sylvestrimale* and *D. saxonica* using stepwise discriminate function analysis of the coordinates of 17 wing vein points (Fig. 3, Table 1).

Not only have the distances between points been used for traditional and geometric morphometrics, but the angles described by wing veins, and some indices based on the points have also been calculated and used as characters for species discrimination (Tüzün, 2009; Kozmus et al., 2011). Also Alexander (1991) used two wing vein angles in his cladistic analysis of the genus *Apis*. Tüzün (2009) used wing vein angles to discriminate between and 30 species of wasps from different families. He used a combination of traditional and geometric morphometric techinques. He used 20 points (see Fig. 2, Table 1) and an additional four points (not shown here) on some species, measured the distance between *all* combinations of points and calculated vein length ratios (Tüzün, 2009). All possible combinations and ratios were calculated and also all angles between points were calculated, yielding a table of 77 different angle and ratio values for all species (Tüzün, 2009). One focus of his study was to differentiate between three *Sphex* species *S. maxillosus, S. flavipennis* and *S. pruniosus*. He used stepwise discriminant fuction analysis and found that the three species were unambiguously separated by this method (Tüzün, 2009). He measured 27 more wasp species and entered the data into a database. He wrote a computer program to compare an unknown specimen with those in the database by calculating:

Total Angle Variation = |Angle 1 $_{[\text{unknown species}]}$-Angle 1$_{[\text{species found in the database}]}$ |+|Angle 2 $_{[\text{unknown species}]}$-Angle 2$_{[\text{species found in the database}]}$ |+…+ etc., and

Total length Variation = |1-Length 1$_{[\text{unknown species}]}$/ -Length 1$_{[\text{species found in the database}]}$ |+ |1-Length 2$_{[\text{unknown species}]}$/ -Length 2$_{[\text{species found in the database}]}$ |+…+ etc. (Tüzün, 2009).

The lower the value the higher the probability of a correct identification. Some examples are given in Table 2 which is extracted from Table 6 of Tüzün (2009). He also calculated a Similarity coefficient = (1/A x R) x K, where A = sum of the differences in wing angles, and R = sum of differences among the ratios of wing veins, and K = a constant (Tüzün, 2009.)

Pre diagnosed species	Species estimated by the program	Sum of differences in wing angles (A)	Sum of differences among the ratios of wing veins (R)	Result: Similarity coefficient
Vespa orientalis	*Vespa orientalis*	**19.873**	**2.111**	**23.8**
	Vespa crabro	51.714	2.856	6.8
	Vespa bicolor	55.962	3.368	5.3
Sphex rufocinctus	*Sphex rufocinctus*	**60.036**	**0.495**	**33.6**
	Sphex maxillatus	87.548	2.459	4.6
	Myzina tripunctata	78.030	2.001	6.4
Eumenes dubius cyranaius	*Eumenes dubius cyranaius*	**16.048**	**0.171**	**364.4**
	Eumenes coronatus detensus	33.398	3.193	9.4
	Eumenes pomiformis	60.840	4.483	3.7

Table 2. Some examples of the identification of wasp species according to wing morphometric values. (Modified from Tüzün (2009)).

His methods are clearly very successful at discriminating between wasp species, at least those represented in his data base. Kozmus et al. (2011) used eight lengths, 17 wing angles and five indices calculated from 19 points (Fig. 1, Table 1) for a total of 37 characters, and measured 530 queens and workers from 18 European species of bumble bees. They did discriminant analysis based on Mahalanobis distance and from this assigned each specimen to a group. Canonical variates analysis was also performed and used to calculated three variables to separate the species into groups (Kozmus et al, 2011). They were able to correctly assign 97% of the bumble bees to the correct species, an in 13 species all the bees were correctly assigned (Kozmus et al, 2011). They found that three characters were particularly informative, based on high R^2 (explained variability) from an ANOVA. These were angle J16, A4 and discoidal shift (Dis D). Angle A4 is the angle described by the points (Fig. 1) 9, 7, 5 (where the vertex is denoted by the second number in the series and then first and last are the end points of line segments), J16 is the angle described by the points 3, 11, 26 and Dis D that between 1, 2, 7 (Kozmus et al, 2011). The R^2 were 63.82%, 61.91% and 60.30% respectively. This particular technique obviously holds great promise for identification and discrimination of *Bombus* species and groups.

The discussion of combined traditional morphometrics and genetic studies (e.g. Aytekin et al., 2003; Owen et al., 2010) will be left until section 4, below.

2.4 Geometric morphometrics

As mentioned earlier, the development of geometric morphometrics has led to the analysis of shape by removing the confounding effects of size. It encompasses a variety of multivariate statistical techniques for the analysis of Cartesian coordinates. These coordinates are usually (but do not have to be) based on point locations called *landmarks* (Slice et al., 2009). The studies which I will discuss here are based on landmarks so the specific suite of techniques used is referred to as *landmark based geometric morphometrics* (Adams et al., 2004). Since it is crucial to understand exactly how landmarks are defined, I have taken the definition directly from Slice et al. (2009):

"**landmark** - A specific point on a biological form or image of a form located according to some rule. Landmarks with the same name, homologues in the purely semantic sense, are presumed to correspond in some sensible way over the forms of a data set.

Type I landmark - A mathematical point whose claimed homology from case to case is supported by the strongest evidence, such as a local pattern of juxtaposition of tissue types or a small patch of some unusual histology.

Type II landmark - A mathematical point whose claimed homology from case to case is supported only by geometric, not histological, evidence: for instance, the sharpest curvature of a tooth."

(There are also Type III landmarks which do not concern us here). It is obvious that vein intersection points on insect wings are ideal Type I landmarks (Figs. 1, 2, 3,4), although they will not be homologous between relatively distantly related taxa (e.g. wasps and bees). It is better to use Type I landmarks and not Type II landmarks (e.g. wingtips) for evolutionary and developmental studies (Aytekin et al., 2007). Therefore differences between wings, either right and left ones of an individual, or differences between species can be analyzed using the Cartesian coordinates of landmarks as the data. The analysis proceeds by removing non-shape variation. This is variation in orientation, position and scale (Adams et al., 2004). There are a number of superimposition methods developed to remove the non-shape variation, but the *Generalized Procrustes analysis* (or just Procrustes analysis) has become widely used. Procrustes analysis is an optimization technique which superimposes landmark configurations using least-squares estimates for translation and rotation parameters (Adams et al., 2004). After superimposition the deformation or "warping" in shape of each individual from a consensus form is given by partial warp scores (Adams et al., 2004; Aytekin et al., 2007). The partial warp scores can be analysed statistically to compare variation in shape within and between populations. Relative warp analysis is a principal component analysis of the partial warps (Adams et al., 2004). The thin-plate spline is used to plot the deformations them on a grid.

Landmark based geometric morphometrics and Procrustes methods have been used in a wide variety of studies over a wide range of taxa (Adams et al., 2004). Three applications of relevance here are (1) allometry of shape, (2) fluctuating asymmetry, and (3) taxonomy and classification.

Allometry of shape was detected by Klingenberg et al. (2001) in their study of development and fluctuating asymmetry in bumble bees, although it was not the main focus of their investigation. In another arthropod, the Fiddler crab, Rosenberg (1997) analysed shape allometry of the major and minor chilipeds. Studies of fluctuating asymmetry will be discussed in section 3, below. Here I will review a few selected studies of Hymenoptera using landmark based geometric morphometrics.

a. Bumble bees: As will be discussed in section 4, there are many taxonomic problems involving the exact status of species in some subgenera of bumble bees (*Bombus*). Aytekin et al. (2007) used landmark based geometric morphometrics to resolve some taxonomic problems in the subgenus *Sibiricobombus*. In particular the specific of *B. vorticus* and *B. niveatus* has been questioned (Williams, 1998). They collected 52 males from six species representing three subgenera (see Table 3).

Species		**Bending energy (10^{-5})**	
	n	Front-wing	Hind-wing
B. (Sibricobombus) niveatus	26	3369	1121
B. (Sibricobombus) vorticosus	6	3850	1117
B. (Sibricobombus) sulfureus	3	4004	1299
B. (Mendacibombus) handlirchianus	6	5073	1816
B. (Melanobombus) erzurumensis	3	2087	289
B. (Melanobombus) incertus	8	294	61

Table 3. The six species of bumble bees collected by Aytekin et al. (2007). Also given are the sample sizes (*n*) and the bending energies for the front- and hind-wings, calculated from the thin-plate-spline. Modified from Aytekin et al. (2007).

Principal component analysis clearly separated all species except *B. vorticus* and *B. niveatus*, also there was no significant difference in size between these two species although all others could be separated by size Aytekin et al. (2007). The bending energies, calculated from the thin-plate-spline showed some difference in the front-wing between *B. vorticus* and *B. niveatus*, but were remarkably similar for the hind-wings (Table 3). They concluded that there were no significant morphological differences between these two taxa and they should be considered conspecific. Aytekin et al. (2007) also concluded that landmark based geometric morphometrics was a powerful method for resolving taxonomic problems in bumble bees and that venation shape may be an important factor for the mechanics of bumble bee flight. Although it must be realised that the bending energies from the thin-plate spline (Table 3) do not represent actual bending energies of a real bumble bee wing, but that they may nevertheless reflect some real mechanical differences between species. Shih et al. (2010) analysed the patterns of wing venation in extinct (fossil) and living pelecinid wasps and identified an "X" pattern of venation in the forewing which evolved and was maintained in some lineages. They suggest that this pattern could have provided a stronger wing structure and led to better flight performance for the larger species (Shih et al., 2010).

b. *Apis mellifera*: Francoy et al. (2009) used geometric morphometrics and an Automated Bee Identification System (ABIS) to examine changes in morphology of an Africanized honeybee population in Brazil 34 years after the African bee swarms escaped. This is interesting because it compare bees collected from 1965 – 1968 with those collected in 2002 at the same location (Francoy et al., 2009). In 1957 swarms of 26 colonies of the African honeybee *Apis mellifera scutellata* escaped in Brazil and hybrized with the previously introduced European honeybee races (Francoy et al., 2009). These Africanized bees have since spread throughout South America, Central America and into the USA by 1990 and are now found as far north as Nevada (Francoy et al., 2009). In 2002 Francoy et al. (2009) collected samples of five workers from 10 colonies from Ribeirão Preto, about 150 km from the original place of introduction of *A. mellifera scutellata*. They measured the right front wing of these specimens, the bees collected from 1965-1968 in the same location and also specimens of *A. mellifera scutellata, A. mellifera carnica, A. mellifera mellifera,* and *A. mellifera lingustica*. They used the standard 19 landmarks on the honeybee wing (Fig. 1, Table 1) and carried out two analyses: (1) geometric morphometrics was done using a Procrustes superimposition followed by the calculation of the relative warps and then a discriminant analysis. Mahalanobis distances, D^2, were also calculated (Table 4) and a dendrogram was plotted using these values (Francoy et al., 2009); (2) ABIS performs an automated analysis of images of honeybee forewings. It analyses the venation pattern and then uses various statistical techniques either linear discriminant analysis or a more powerful Kernal discriminant analysis which allows species and subspecies identification (Francoy et al., 2009). The system has to be "trained" with at least 20 specimens of each group (Francoy et al., 2009).

	RP - 1968	RP - 2002	*A. mellifera scutellata*	*A. mellifera mellifera*	*A. mellifera carnica*	*A. mellifera lingustica*
RP - 1968	-	12.43	12.40	21.60	32.98	29.83
RP - 2002		-	15.14	24.18	37.68	34.04
A. mellifera scutellata			-	22.47	27.08	23.65
A. mellifera mellifera				-	34.54	29.68
A. mellifera carnica					-	9.32
A. mellifera lingustica						-

Table 4. Mahalanobis distances, D^2, between the centroids of the Apis mellifera groups calculated through relative warp analysis (modified from Francoy et al., 2009). RP = Ribeirão Preto populations.

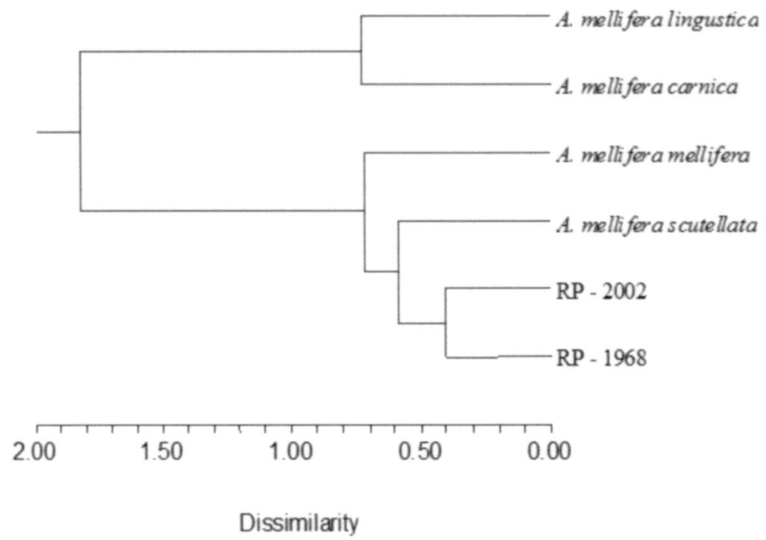

Fig. 5. A dendrogram produced by the Unweighted Pair-Group method (UPGMA) using the Mahalanobis distances in Table 4. Francoy et al. (2009) used the neighbour-joining tree method to produce a very similar dendrogram. Note that RP = Ribeirão Preto populations.

From Table 4, from which a dendrogram is constructed (Fig. 5) it is clear that Africanized bees resemble the African race more than they do the European races. Also it is interesting to see that there have been some morphological changes in the Africanized bees in Brazil over the 34 years since the hybridization event. The 1968 and 2002 Ribeirão Preto populations are clearly distinct with a Mahalanobis distance of 12.43 (Table 4, Fig. 5). The ABIS gave essentially the same results.

As mentioned earlier Tüzün (2009) used both traditional and geometric morphometric techniques. Clustering of the relative warps also separated the three *Sphex* species very well.

Two more studies are of interest as they show slightly different applications of the techniques. Pretorius (2005) used standard geometric morphometrics to examine wing shape dimorphism between male and female wasps in the genus *Tachysphex*. He used 24 species in this genus and measured 15 landmarks (Fig. 2, Table 1). He did find small but definite differences in the shapes of the wings between the sexes and cautioned that in an analysis of a genus only one of the sexes should be used as small-scale differences, may in some cases influence the results (Pretorius, 2005).

Baylac et al. (2003) studied two closely related species of Bracoinid parasitoids. The two species *Bassus tumidulus* and *B. tegularis* had been synonymised and then subsequently split. Baylac et al. (2003) used geometric morphometrics of wing venation to study this problem but they also used some aspects of pattern analysis. Pattern analysis involves statistical techniques such as kernel density estimates and Gaussian mixture analysis (Baylac et al., 2003). Baylac et al. (2003) measured 15 landmarks on the wing (Fig. 4, Table 1) and found that both methods did separate the species into two definite morphological groups.

It is often useful and informative to combine traditional and geometric morphometric techniques and other methods (Fruciano, et al., 2011; Tüzün, 2009; Baylac et al., 2003). For instance Fruciano et al. (2011) point out that it may be necessary to use traditional techniques to allow a comparison with earlier results in the literature.

3. Fluctuating asymmetry

Most animals are bilaterally symmetrical, with paired internal organs and paired appendages. However the symmetry is often not exact or "perfect". There are two general classes of asymmetry; conspicuous and subtle; conspicuous asymmetries are very obvious, for example the extreme difference in size between right and left claws in some crabs, e.g. Fiddler Crabs. However many animals exhibit less obvious types of asymmetry which can only be quantified in a *sample* of individuals, and thus statistical methods must be used to analyze it (Palmer, 1994). Measurements are made on a structure on the right (R) and left (L) sides of each individual in the sample and an index of asymmetry is then calculated. Three types of asymmetry can occur: (1) fluctuating asymmetry (FA), with a normal distribution of R- L values around a mean of zero, (2) directional asymmetry (DA) where the mean of one side is almost always greater than that of the other, and (3) antisymmetry where there is a difference between the two sides but it cannot be predicted which will show the greater value, so giving a broad-peaked or bimodal distribution of R-L values about a mean of zero (Palmer & Strobeck 1986).

Developmental stability (DS) is defined as "the ability of an organism to buffer development against genetic or environmental perturbation" (Clarke, 1997). For instance, populations undergoing decline are likely exposed to environmental and genetic stresses which may cause developmental instability (DI) of individuals (Parsons 1990, Milankov et al. 2010). This DI is often manifest by deviations from bilateral symmetry (Palmer 1994). Insect wing venation characters are ideal for assessing FA and environmental stress. Fluctuating asymmetry (FA), where the differences between right and left sides follow a normal distribution, should reflect perturbations from perfect bilaterally symmetrical development and thus serve as a measure of the stresses experienced by an individual during its development (Palmer & Strobeck 1986). In turn it can be used as an epigenetic measure of stress in natural populations (Parsons 1990). Therefore the estimation of FA could be an important indicator of the "health" of species and help guide decisions regarding conservation. Recently much attention has been paid to the significant contraction of the distributions, and the decline in the abundance, of some bumble bee species in North America and Europe (Evans et al. 2008; Goulson et al. 2008). In North America *Bombus affinis* Cresson, *B. terricola* Kirby and *B. occidentalis* Greene have all disappeared from significant parts of their historic ranges (Colla and Packer 2008; Evans et al. 2008, Grixti et al. 2009; Cameron et al. 2011). If FA is a good predictor of stress then we would predict higher levels of FA in species undergoing decline than stable species.

3.1 Developmental stability, fluctuating asymmetry and quantitative genetic variation

Here I clarify the relationship between developmental stability, fluctuating asymmetry and quantitative genetic variation in the Hymenoptera. Hymenoptera (ants, bees and wasps), have what is known as a *haplodiploid* genetic system. This means that females (queens and workers in eusocial species), like most animals, are derived from fertilized eggs while males arise from *unfertilized* eggs. Thus males are *haploid* (n) inheriting only one member of each pair of chromosomes, those from their mother, whereas females are *diploid* ($2n$) having both members of each pair of chromosomes. Formally the system of inheritance follows the pattern of X-linked inheritance in organism in which both sexes are diploid. Many aspects of the genetics of X-linked or haplodiploid genes are different from that of autosomal genes, including the expression of quantitative or polygenic characters. One consequence is that for a genetically determined quantitative trait males are likely to be more variable than females (Eickwort, 1969; Owen, 1989). The variances are derived on the assumption of dosage compensation of genotypic values in males (Fig. 6), and we must distinguish between mean within-family (or within-colony) variances and population variances. Consider a single gene locus with alleles A_1 and A_2 at frequencies p and q respectively, and let the genotypes take the genotypic values shown in figure 6.

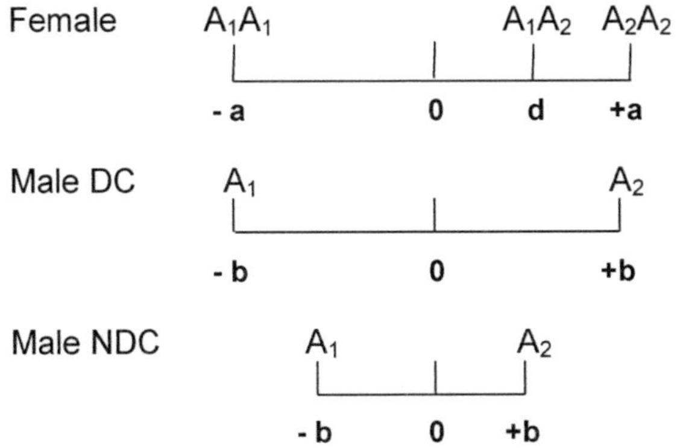

Fig. 6. Arbitrarily assigned genotypic values of a quantitative trait at an X-inked or haplodiploid locus, showing the difference between dosage compensation (Male DC) and no dosage compensation (Male NDC) of male genotypic values.

The well-known population variances (Owen, 1989) are,

Females: $$V_f = 2pqa^2 + (2pqd)^2$$

$$= V_{Af} + V_{Df}$$

Males: $$V_m = V_{Am} = 4pqb^2$$

Where the average effect $a = a+d(q-p)$,and $V_{Af}$and V_{Df} are the female additive and dominance variance components, respectively. Note that the male genotypic variance, V_m consists solely of an additive component, V_m. The corresponding mean within-colony variances are,

Females: $$\bar{V}_f = \tfrac{1}{2}pq[\alpha^2 + 2ad(q-p) + d^2]$$

Males: $$\bar{V}_m = 2pqb^2$$

It is assumed that the male and female offspring are full-siblings. Thus we can see that with (i) no dominance in females (d = 0), and (ii) dosage compensation in males (a = b), then for the population variances,

$$V_m = 2V_f$$

and for the mean within-colony variances,

$$\bar{V}_m = 4\bar{V}_f$$

Thus males are predicted to more variable than females, and this will generally be the case unless there is complete dominance in females and only then when the allele frequency q>0.62 (Owen, 1989). However if there is no dosage compensation then females will be more variable, i.e. $V_f > V_m$, except with intermediate dominance when q<0.16 and with complete dominance when q<0.21 (Owen, 1989). This differential variability has no relationship *per se* with developmental stability and FA, it is just the result of the different ploidy levels in males and females. However if genome wide heterozygosity promotes DS then we would expect haploid males to show more DI and FA than their diploid counterparts, thus haplodiploid organism are good models with which to partition the effects of heterozygosity and ploidy on DI and FA (Clarke, 1997; Smith et al. 1997). The prediction is that, due to the absence of heterozgosity, the haploid males will show higher FA than the diploid females.

3.2 Differential morphological variation between the sexes in the hymenoptera

Males in many species of Hymenoptera, in accordance with quantitative genetic theory are indeed more variable in morphological characters than females, although this is not always the case. For comparisons of eusocial species it is important to compare reproductives, i.e. males with queens and not workers. In the Hymenoptera worker size variation is great and due to many different factors (Wilson, 1971). Eickwort (1969) in her multivariate morphometric study of *Polistes exclamans* found that males were more variables than queens in the characters used. In addition to wing measurements (see Fig. 2 and Table 1) she also used six other morphological characters (number of hamuli, distance between the most distal and proximal hamular sockets, mesoscutal width and length, distance between compound eyes, head width). She sampled 19 nests and calculated generalized variances (D) to compare mean within-colony (nest) variances of males and females. Variance-covariance matrices were calculated across all the characters for males and females in each nest and the determinants of these matrices were defined as the generalized variances. Males were more variable than queens (P<0.01), and she noted that even the largest generalized variance for queens (0.000006020) was smaller than the smallest generalized variance of any group of males (0.000075343) (Eickwort, 1969).

Univariate studies of differential variability in the Hymenoptera are relatively common. I examined variation of radial cell length (distance between points 1-2, Fig. 1) in the bumble bee *Bombus rufocinctus.* A sample of 787 young queens from 38 laboratory reared colonies and a sample of 680 males from 38 colonies were measured. The males, with a coefficient of variation (CV) of 5.75% were significantly more variable (P<0.01) than the females (CV=3.98%). There were also significant intraclass correlations between male (t_m = 0.553) and young queen (t_f = 0.435) offspring, indicating a considerable degree of phenotypic resemblance between bees of the same caste within each colony (Owen, 1989). Heritability as estimated from offspring-parent regression (0.20 ± 0.19 for queens and 0.47 ± 0.38 for males) was significantly lower that than estimated from the intraclass correlations, suggesting that environmental variation is of the same order of magnitude as additive genetic variation (Owen, 1989).

In other Hymenoptera environmental variation clearly is the most important determinant of phenotypic variation. Owen & McCorquodale (1994) examined variation and heritability of body size and postdiapause development time in the leafcutter bee, *Megachile rotundata*. The bees were from a domesticated population and the nests were in pine blocks (12 cm long) with about 500 standard nest tubes 5 mm in diameter (Richards, 1984). Head widths of offspring from total of 200 nests was measured and the frequency distribution is shown in figure 7.

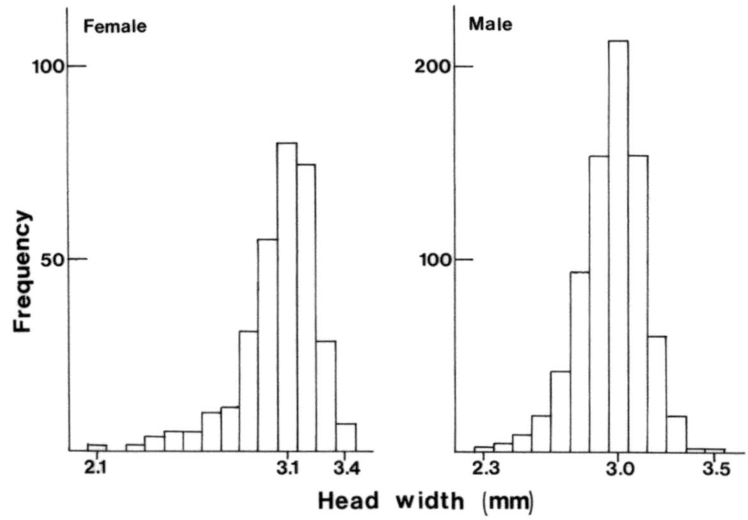

Fig. 7. Frequency distributions of head width in a sample of female and male offspring from 200 nests of Megachile rotundata (modified from Owen & McCorquodale, 1994).

Female offspring (n=312) from 151 of these nests had a mean head width (±SEM) = 3.5 ± 0.011 mm, and male offspring (n=769) from 172 of the nests were significantly smaller ($P<0.00001$) with mean head width = 2.96 ±0.006 mm. Interestingly the males also were *less* variable with a CV=5.65% as compared to females with a CV of 6.58%. Heritability of head width, estimated from offspring-parent regression, was not significantly different from zero and was considerably lower than that obtained from the intraclass correlation coefficient. Because intraclass correlations can be inflated by environmental variation, the difference between these two estimates implies that in this case maternal effects are the most important determinant of head width (Owen & McCorquodale, 1994). O'Neill et al. (2010) found that in feral populations of *M. rotundata* offspring size (head width) was generally positively related to tunnel diameter. Again, as Owen & McCorquodale (1994) did, they found offspring within families were more similar to each other than to bees in other families.

O'Neill et al. (2010) concluded that the most important maternal effect which probably accounted for this was the amount of provision provided by the mother. In *M. rotundata* and other leafcutter bees there is low if any, genetic variation for body size, and environmental factors are the major causes of variation in males and females (Owen & McCorquodale, 1994).

In mass provisioning wasps the size of the offspring is determined to a great extent by the mass of provision they receive (Hastings et al., 2008), and so the variation in offspring size would also reflect the variation in mass provision size. Therefore we would expect genetic sources of variance to be quite weak in comparison to environmental causes. This is illustrated well by the cicada killer wasp, *Specius speciosus* (Hastings et al., 2008; 2010). Hastings et al. (2010) studied two populations of Eastern cicada killer wasp in northern Florida, and they measured wet mass (mg) and right wing length (mm) of males and females. Both males and females were significantly larger and heavier at St. Johns than at Newberry, but in both populations males were less variable than females. The CV's calculated from Table 1 in Hastings et al. (2010) are: St. Johns, males CV=0.08%, females CV=0.10%; Newberry, males CV=0.05%, females CV=0.07%. There is a very interesting relationship between body size of the wasps and their prey. *S. speciosus* females provide male offspring with usually a single cicada, and each female offspring with usually two cicadas irrespective of prey size (Hastings et al., 2008). Hastings et al. (2008) sampled the wasps and cicadas from 12 different locations in 10 states in the USA, and they found a significant correlation between wasp size and the mean local cicada mass. However they did find the two locations in Florida (St. Johns and Newberry) where the pattern did not hold. Hastings et al. (2010) found that in these populations female wasps exhibited prey selection by size. Small wasps only collected small cicadas and large wasps only collected large cicadas. The small wasps probably cannot carry the large cicadas but the large wasps, which could carry the small ones, select only the larger sizes (Hastings et al., 2010).

3.3 Fluctuating asymmetry in haplodiploids

There are relatively few studies of FA and DI in Hymenoptera and haplodiploid organisms. Some studies have used traditional morphometric methods while others have employed geometric morphometric techniques. Clarke's (1997) study was to test the hypothesis that haploid males should show greater DI than the diploid females, as manifest by larger FA in males. He used a combination of morphometric (wing vein lengths; the details were unspecified) and meristic (number of humuli) in six taxa of Hymenoptera; two races of *Apis mellifera* (*capensis* and *scutellata*), *A. cerana*, *Trichocolletes affenutus* (Colletidae) , *Vespula germanica* (Vespidae) and *Solenopsis invicta* (Formicidae). He also assessed two haplodiploid thrip (Tysanoptera) species, *Haplothrips angustus* and *H. froggatti*, the measure used was the number of duplicated cilia along the posterior margin of the forewing. Clarke (1997) calculated mean asymmetry values for each character in each sex, and tested the difference between sexes using single classification and multivariate analysis of variance. Clarke's (1997) did not find any consistent pattern and his conclusion was that, as a whole, haploid males are no more asymmetric than diploid females. Of the 60 direct comparisons made using univariate ANOVA only 8% showed the haploid males to be more variable than the females and only 3% showed the reverse, and the other comparisons showed no significant difference. Clarke (1997) found no significant difference in asymmetry between males and females in the two *Haplothrips* species. Crespi & Vanderkist (1997) measured FA in the thrip, *Oncothrips tepperi* also to test the hypothesis of higher FA in males, and also to compare FA in functional and vestigial traits. The latter should exhibit higher FA than the former due to relaxion of selection for functionality (Crespi & Vanderkist, 1997). They measured fore femora lengths of soldier and disperser morphs, and wing lengths of dispersers (functional traits), and wing length of soldiers (vestigial trait). Analysis was done following the methods of Palmer (1994). They found complex interactions between sex, caste and FA, namely that for wings FA was higher in female soldiers that in male soldiers, but in dispersers males had the higher FA. For the femora males and females did not differ in FA in either morph. Crespi & Vanderkist (1997) concluded that there was no consistently higher FA in males than females, but that vestigial traits did show higher FA than functional traits.

Silva et al. (2009) estimated FA in two species of Euglossine bees in Brazil to assess the effects of climatic and anthropogenic stresses on these bee populations. They collected 60 males of each species, 30 from the forest border and 30 from the interior of the forest, and half were collected during the hot, wet season and the other half during the cold, dry season (Silva et al., 2009). Four measurements (M1, M2, M3, M4, see Fig. 1, Table 1) were made on both wings of each individual, and for each measurement FA was calculated. A general body size index was obtained from a principal component analysis of measurements M 1-3, and then the transformed FA and size index data were analysed using ANOVA (Silva et al., 2009). There were no differences in FA for the four characters between areas and seasons in *Eulaema nigrita*, however in *Euglossa pleostica*, they found significant greater FA of M3 in bees collected in the hot and wet season than those collected in the cold and dry season. Silva et al. (2009) concluded that this species was responding to increased environmental stress in the hot, wet season.

The last two studies of FA that I will discuss used geometric morphometric methods. Smith et al. (1997) were interested in partitioning out the effects of ploidy and hybridization on levels of FA in *A. mellifera*. They used the coordinates of 19 points (see Fig. 1 and Table 1) on the forewings of ten workers and five males (drones) from each of 27 hives. The coordinates were digitized and subject to a Procrustes analysis of asymmetry (Smith et al., 1997). The specialized analysis described by Smith et al. (1997) produces a measure of asymmetry, A^2, for each specimen, then the mean A^2 of a series of specimens is decomposed into one term for FA and another term for directional asymmetry (DA). Smith et al. (1997) found that across all populations total asymmetry was significantly greater (one-way ANOVA, $P<0.001$)for haploid males than for diploid females, however they were surprised to find that most of the asymmetry was not due to FA but was directional asymmetry (Table 5).

	All bees	Females	Males
n=	377	261	116
Total squared asymmetry	7.29	6.34	9.42
Directional squared asymmetry	3.61	2.89	6.31
Fluctuating squared asymmetry	3.68	3.45	3.11

Table 5. Partitioning of total squared asymmetry in *Apis mellifera* into directional and fluctuating components. Note all entries are x 10^4. Modified from Smith et al. (1997).

Smith et al. (1997) concluded that perhaps DA was more common than previously thought. Klingenberg et al. (2001) examined FA and variation among individuals in the forewings and hindwings of bumble bees as part of an investigation of developmental modularity. The fore- and the hindwings develop from separate imaginal discs and so are expected to be independent developmental modules (Klingenberg et al., 2001). Klingenberg et al. (2001) predicted that patterns of variation among individuals should be similar to the patterns of FA within each wing, and that individual variation between fore-and hindwings will co-vary (depending on how much they really are independent modules), but that FA will be independent between them. They measured 13 points on the forewings (see Fig. 1 and Table 1) and six on the hindwings of worker bees. They used laboratory reared bumble bee colonies and subject sets of colonies to three treatments which consisted of providing a flow of air through the colonies, two with different concentrations of CO_2, 10% and 5%, and one a control treatment with just air (ultimately they only used the control and 5% treatments). Klingenberg et al. (2001) used geometric morphometric and Procrustes methods to characterize size and shape variation in fore- and hindwings separately. They found that the major pattern of variation within each wing was the coordinated shifts in sets of landmarks over the entire wing. This suggests that each wing is a developmental module which is not further subdivided into smaller domains (Klingenberg et al., 2001). As a consequence they also concluded that any small perturbations causing FA are transmitted throughout the entire wing, affecting all landmarks. Since shape asymmetry co-varied only between fore- and hindwings in the CO_2 treatment Klingenberg et al. (2001) concluded that the developmental interactions between wings are probably related to gas exchange.

4. Taxonomic and systematic problems: Concordance between genetic and morphometric approaches in bumble bees

Here I will discuss the use of combined genetic and morphometric approaches to resolve taxonomic problems, with examples from bumble bees. Bumble bees (tribe Bombini) form a well-defined monophyletic group containing a relatively small number of species (239 according to Williams 1998), thus it may seem surprising that bumble bees pose many taxonomic and systematic problems. At the specific level the taxonomic status of closely related taxa is often unclear and subject to contradictory interpretations. Bumble bees are relatively quite invariant or 'monotonous' morphologically compared to other bees (Michener 2000), but many species show considerable pile colour variation. Some of this has a simple (Owen & Plowright 1980) or relatively simple (Owen & Plowright 1988) genetic basis, but most variation is continuous and probably polygenic in nature (Stephen 1957), and to complicate matters further, considerable convergence in colour pattern, often between distantly related species also occurs (Plowright & Owen 1980). The root of the problem is that traditional taxonomic approaches are limited when applied to bumble bees. Genetic and statistical methods must be used to understand processes of speciation in *Bombus*. For example, Scholl *et al*. (1990) found that *B. moderatus* differed from *B. lucorum* at 3 out of 26 enzyme-gene loci, with the electromorphs exhibiting fixed differences in each species. Again, Scholl *et al.* (1992) found fixed electrophoretic differences between *B. auricomus* and *B. nevadensis* at 5 out of 18 enzyme loci. In both cases the authors suggested the return to the original specific designations. A powerful approach, which has been very successful in resolving some of these problems, is to combine genetics and morphometrics.

Aytekin et al. (2003) combined these approaches to elucidate the relationship between two subspecies of *Bombus terrestris*. In the eastern Mediterranean region two subspecies have been recognized, *B. terrestris dalmatinus* from the Balkans and surrounding areas; and *B. t. lucoformis* from Anatolia (Aytekin et al., 2003). Aytekin et al. (2003) sampled 157 specimens of queens and workers from Bulgaria, Greece and Turkey. They assessed allozyme variation by using six enzyme systems and morphometric variation by using 28 morphological characters. Of the morphological characters employed 13 were distances measured between points, eight on the front wing (Fig. 1, Table 1) and five on the hindwing. They found that the allozymes exhibited very little variation and the electromorphs appeared to be fixed in all populations, and both taxa were monomorphic in all loci scored (Aytekin et al., 2003.). They found no heterozygotes or different electromorphs, except *B. t. lucoformis* found in the Ankara region had two alleles for malic enzyme (*Me*) with electrophoretic mobilities of 100 and 102. The morphological characters were analysed by multigroup discriminant function analysis (canonical variates CANOVAR) and principal component analysis (PCA), and also failed to separate the two groups, so (Aytekin et al., 2003) concluded that there was not enough of a difference between *lucoformis* and *dalmatinus* to warrant separate sub-species status. I will now discuss two examples of some of my own work in more detail.

4.1 *B. melanopygus/ B. edwardsii*

Owen et al. (2010) examined the relationship between the two nominate taxa *B. melanopygus* Nylander, and *B. edwardsii* Cresson, using a combination of genetic and morphometric analyses. Traditionally there was absolutely no question that these taxa represented two distinct species (Stephen 1957; Milliron 1971) since the bees differ dramatically in the colour of the abdominal terga two and three, these being ferruginous (or red) in *B. melanopygus* and black in *B. edwardsii*, although other morphological differences between the two are minor (Stephen 1957; Owen et al. 2010). Moreover, the distributions have relatively little overlap. *B. edwardsii* occurs throughout California and just into neighbouring Nevada, while *B. melanopygus* extends north through Oregon, Washington, British Columbia, Alaska, east into Alberta, Saskatchewan, and across northern Canada possibly to Labrador (Stephen 1957; Laverty and Harder 1988). They are sympatric only in southern Oregon and northern California (Stephen 1957). However, the taxonomic status of these bees was called into question when Owen & Plowright (1980) reared colonies from queens collected in the area of sympatry. They discovered that pile coloration was due a single, biallelic Mendelian gene, with the red (*R*) allele dominant to the black (*r*). Also, the observed numbers of queen genotypes and colony types at each collection location conformed to those expected under Hardy-Weinberg equilibrium. This suggested that the two taxa are in fact conspecific, in which case there is a gene frequency cline running from north to south where the red allele is completely replaced by the black allele over a distance of about 600 km (Owen & Plowright 1980; Owen 1986). Although this genetic evidence is compelling, because the bees were only collected from the region where both alleles are present, it still leaves open the logical possibility that *B. edwardsii* is the dimorphic species and *B. melanopygus* exists as a separate, northern species.

Owen et al. (2010) showed that both enzyme electrophoresis and wing morphometrics do unambiguously distinguish between these two species. Allozyme electrophoresis can be useful for distinguishing closely related species. If there are fixed differences, or large gene frequency difference between two taxa then this would strongly suggest either complete, or a very high degree of, reproductive isolation. Conversely, if two taxa have identical allozyme profiles, then this would strongly suggest conspecificity (see above Aytekin et al., 2003); however it cannot of course *prove* it. Similarly, morphometric analysis of wing venation patterns has also proved to be very successful for differentiating between bumble bee species as discussed earlier. Owen et al. (2010) included in their analysis a closely related species, *B. sylvicola* with which *B. melanopygus* is sympatric in Alberta. This was to verify that the techniques they used were sensitive enough to correctly discriminate closely related species if real differences do exist. Specimens were collected from Alberta and locations in Oregon and California (Fig. 8) and 113 bees were scored at 16 enzyme-gene loci using horizontal starch gel electrophoresis. For details see Owen et al. (2010). Traditional morphometrics was used and the points measured were a subset of those used by Plowright & Stephen (1973) The distance from 18 to the 13 points shown (Table 1, Fig. 1) was measured (for more details see Owen et al. (2010). Discriminant analysis was done using the statistical software package NCSS (Hintze 1996).Owen et al. (2010) did not standardize the measurements as done by Plowright & Stephen (1973), for two reasons: one was to ensure that any differences between taxa would be *maximized* by the discriminant analysis and the other reason was because size of bumble bee queens is important ecologically (Owen 1988), which might reflect real differences between the species if they exist.

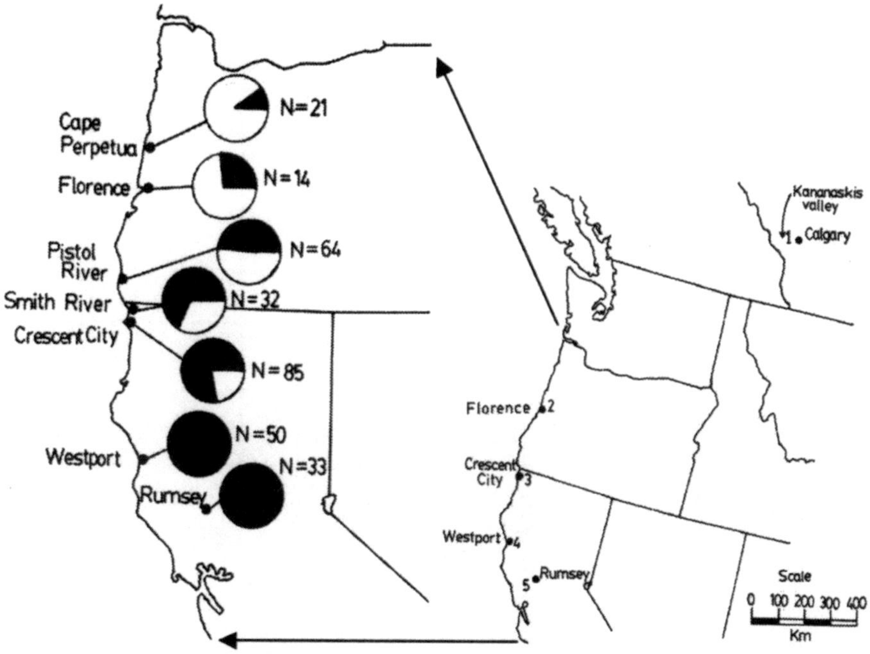

Fig. 8. Collection locations for bees examined electrophoretically and morphometrically. The enlarged section shows the gene frequency cline in *Bombus melanopygus* in Oregon and California. Pie diagrams give the relative frequency of the *R* (red) allele (clear portions) and the *r* (black) allele (shaded portions). The sample size (N) at each location represents the combined total of queen bees collected in1978, 1979, 1980, 1981 and 1988.

All bees had identical electrophoretic mobilities, and were invariant at 11 of the 16 enzyme loci examined. Five loci exhibited either differences between taxa and/or variation within taxa (Table 6). The nominate forms of *sylvicola* and *melanopygus* from Alberta clearly have different electrophoretic profiles (Table 6). The electrophoretic profiles of *melanopygus* and *edwardsii* from all locations were entirely consistent with each other. There was a very small amount of variation present, with heterozygotes being detected at a few locations (Table 6).

What was really interesting was the six bees ("MEL X"), collected in Alberta, that were assigned to *melanopygus* by eye when they were collected but turned out to have an electrophoretic profile inconsistent with that of *melanopygus* but consistent with that of *sylvicola* (Table 6). Going back to the collection records it was found that these bees (plus another three that were not electrophoresed) came from high elevations in the Kananaskis Valley (Fortress Mountain and Highwood Pass) where typical *sylvicola* had been collected. These were later reassigned to *sylvicola* on the basis of the wing morphometric analysis (see below).

	Enzyme electromorph													
	Pgm				*Gpi*		*Idh* (NAD)			*Hk*			*Sdh*	
Taxon*	72/82	82	93	93/100	92/96	96	95	100	102	100	100/105	105	100	105
***B. sylvicola* (*n*=18)**	1	17				18	16	2				18	1	17
"X" (*n*=6)		6				6		6			1	5	6	
"*B. melanopygus*" AB (*n* = 16)			16			16		13	3	16			16	
"*B. melanopygus*" OR/CA (*n*=25)			23	2	1	24		15	10	25			25	
"*B. edwardsii*" OR/CA (*n*=48)			24	2	2	46		35	13	48			48	

* Taxon: *B. sylvicola*; "X" = the bees from Alberta resembling melanopygus, but with an electrophoretic profile inconsistent with the other *melanopygus*; *B. melanopygus* AB = from Alberta; *B. melanopygus* OR/CA = from Oregon and California; *B. edwardsii* OR/CA = from Oregon and California

Table 6. Electrophoresis results for the five enzymes exhibiting either differences between taxa and/or variation within taxa. The other 11 loci were invariant within, and showed no differences between, all taxa. The body of the table gives the number of individual bees of each electromorph. Electromorph mobilities (mm) are standardized relative to those of B. occidentalis (= index 100, Scholl et al. 1990). Modified from Owen et al. (2010).

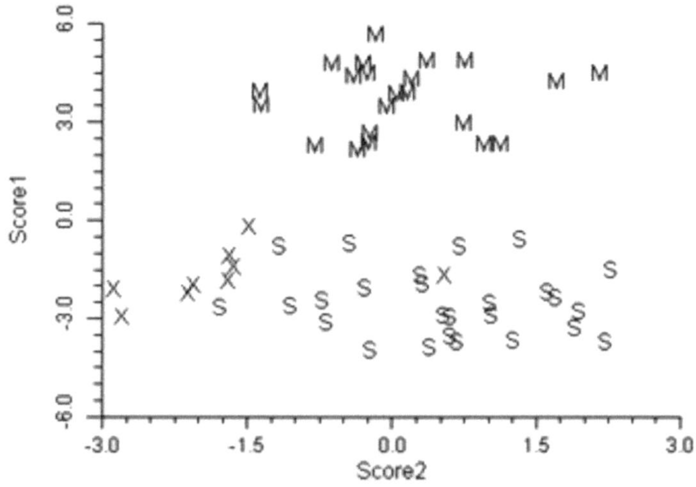

Fig. 9. Plot of the first two Canonical scores for *B. sylvicola* (S), the Alberta *B. melanopygus* (M) and the anomalous Alberta *B. melanopygus* ("Mel X"). Modified from Owen et al. (2010).

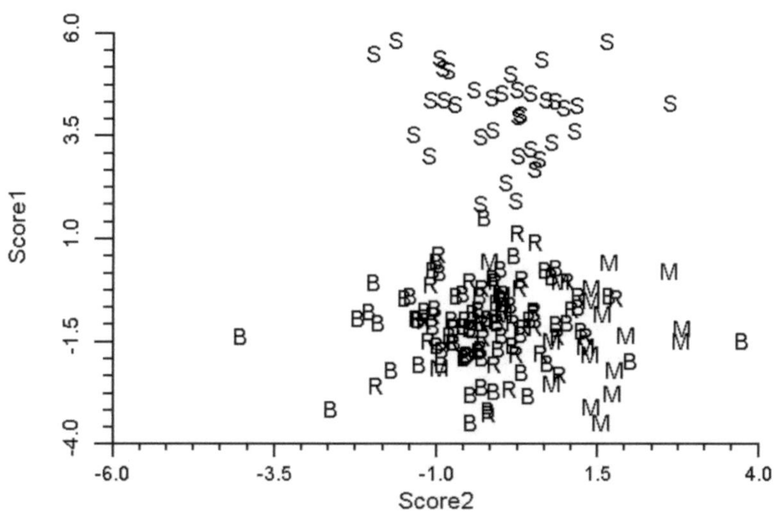

Fig. 10. Plot of the first two Canonical scores for the total data set. S = *B. sylvicola*, R = red "*melanopygus*" from Oregon and California, B = black "*edwardsii*" from Oregon and California, M = *B. melanopygus* from Alberta. Modified from Owen et al. (2010).

The discriminant functions analysis was run three times. Initially only specimens from Alberta were included. This was to verify that the technique could separate closely related species (*melanopygus* and *sylvicola*) in sympatry, and to determine the status of the aberrant *melanopygus* ("MEL X"). In addition to the six "MEL X" bees that were electrophoresed (Table 6) three other queens that were collected on the same dates and at the same locations were reassigned from *melanopygus* and included in the "MEL X" category. The plot of the first two canonical scores is shown in Figure 9. *B. melanopygus* is clearly separated from *B. sylvicola* by the first canonical score. Similarly the "MEL X" bees are obviously distinct from *melanopygus* and are grouped with *sylvicola*. Next, the analysis was run using the complete data set (Figure 10) with the "MEL X" bees now being reclassified as *sylvicola*. Again, *B. sylvicola* is clearly separated by canonical score one, but *melanopygus* and *edwardsii* are not obviously resolved.

Enzyme electrophoresis and wing morphometrics failed to distinguish the nominate species *B. edwardsii* and *B. melanopygus*, yet clearly separated *B. sylvicola* from the latter. This, together with the colour dimorphism genetic data (Owen and Plowright 1980), and the lack of other morphological differences led Owen et al. (2010) to conclude that *melanopygus* and *edwardsii* are conspecific. If *B. melanopygus* is a "good" species, then there is a gene frequency cline for the color dimorphism (Fig. 8).

4.2 *B. occidentalis/B. terricola*

Two other taxa, where the evolutionary status and taxonomic classification are also unclear, are *B. terricola* Kirby and *B. occidentalis* Greene. The basis of this confusion originates with their classification being based primarily on pile colour pattern. Greene's original description of *B. occidentalis* reads "…first four abdominal segments black… "(Franklin 1913). Given that this is the type specimen description, specimens with the first four abdominal segments being black should be considered 'typical' *B. occidentalis.* In contrast, typical *B. terricola* have TIII and TIV that are consistently and clearly defined by complete yellow bands, and lack the large amount of white to cream-coloured pile typical of *B. occidentalis* on TV and TVI. However, in some parts of its distribution including areas of overlap with *B. terricola, B. occidentalis* exhibits considerable pile colour variation with some specimens closely resembling *B. terricola* (Stephen, 1957; Milliron, 1971). The primary ambiguous components of these bees are the complete to incomplete yellow bands on gastral terga III and IV. Nevertheless Stephen (1957) noted that *B. terricola* was "one of the most color stable species in western America" (p. 82) showing little or no variation throughout its range, and that it could be distinguished from *B. occidentalis* in having TII always yellow and TIV black. On this basis many authors have regarded *B. occidentalis* and *B. terricola* to be separate species (Stephen 1957; Thorp et al. 1983). However, Milliron (1971) reduced *B. occidentalis* to subspecific status under *B. terricola,* citing a lack of evident reliable or constant morphological features by which to differentiate specimens in areas of overlap. Milliron (1971) also suggested that these two subspecies most probably interbreed, producing numerous perplexing subspecific hybrids. This is certainly one possible explanation for the rare occurrence of colonies headed by definite *B. occidentalis* queens which produce *B. terricola*-like offspring.

Recently Bertsch et al. (2010) sequenced part (1005 bp) of the mitochondrial cytochrome oxidase subunit I (COI) gene and found a difference of 30 nucleotides between *B. occidentalis* and *B. terricola,* which is significantly larger than that found within a species. On this basis Bertsch et al. (2010) concluded that *B. occidentalis* and *B. terricola* do represent good biological species. They also suggested that to clarify the situation these taxa should be studied in greater detail in their area of contact in British Columbia and southern Alberta.

Whidden (2002) studied sympatric populations of *B. occidentalis* and *B. terricola* in Alberta using randomly amplified polymorphic DNA (RAPD) analysis. For comparison he also analyzed one consubgeneric species, *B. moderatus,* and one non-consubgeneric species *B.*

(*Pyrobombus*) *perplexus*. Ninety two bands using four different PCR primers were generated. Fixed differences occurred between all groups, and individual haplotypes did not occur in more than one taxonomic group, although there was overlap in haplotype components. The corrected average number of pairwise differences of between *B. moderatus* and *B. terricola* and *B. moderatus* and *B. occidentalis* was 6.98 and 5.92 respectively, and that between *B. occidentalis* and *B. terricola* was 5.07 (Table 7).

Species (n)	*B. terricola*	*B. occidentalis*	*B. moderatus*	*B. perplexus*
B. terricola (87)	1.28	6.27	7.91	54.76
B. occidentalis (79)	5.07	1.11	6.77	53.87
B. moderatus (104)	6.98	5.92	0.59	53.55
B. perplexus (54)	53.21	52.41	52.34	1.81

Table 7. Average pairwise differences between and within bumble-bee species.Above diagonal: Average number of pairwise differences between groups (P_iXY).Diagonal elements: Average number of pairwise differences within groups (P_iX). Below diagonal: Corrected average number of pairwise differences (P_iXY-(P_iY)/2). Sample sizes are given in parentheses.

Traditional morphometric analysis was done on some specimens of *B. occidentalis* and *B. terricola* queens collected in 1985 and 1986. The left forewing was removed and measured using the methods of Owen et al. (2010) as described above, and discriminant analysis performed. The classification counts are given in Table 8, and the first and third canonical scores are plotted in Fig. 11.

Species		**Predicted**				
Actual[1]	*occidentalis* 1985	*occidentalis* 1986	*terricola* 1985	*terricola* 1986	Total (n)	% correctly classified
occidentalis 1985	16	15	4	5	40	77.5%
occidentalis 1986	17	39	2	5	63	88.9%
*terricola*1985	2	0	17	4	23	91.3%
terricola 1986	1	3	6	15	25	84.0%
Total	36	57	29	29	151	85.4%

[1] Reduction in classification error due to variables measured = 43.5%.

Table 8. Classification count (actual and predicted) of the *B. occidentalis* and *B. terricola* from 1985 and 1986 using discriminant analysis of wing venation.

The taxa are clearly separated by both the genetic and morphological evidence. The corrected average number of pairwise differences of between *B. moderatus* and *B. terricola* and *B. moderatus* and *B. occidentalis* was 6.98 and 5.92 respectively, and that between *B. occidentalis* and *B. terricola* was 5.07. Therefore since *B. terricola* and *B. occidentalis* are differentiated from each other to the same extent as they are from *B. moderatus*, they should regarded as distinct taxa. Discriminant function analysis of wing morphometric data correctly classified over 85% of the specimens of *B. occidentalis* and *B. terricola*, indicating significant morphological divergence.

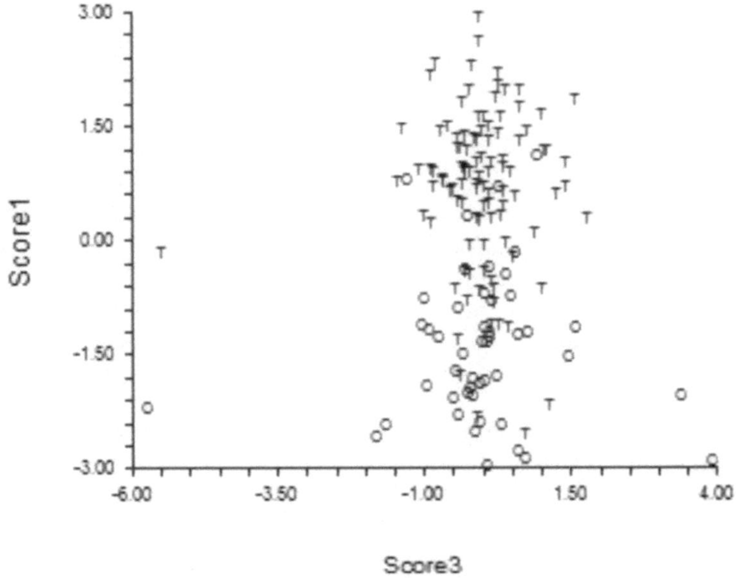

Fig. 11. Plot of the first and third Canonical scores for the 1985 and 1986 specimens of *B. occidentalis* (O) and *B. terricola* (T).

5. Conclusions

Morphometric analysis has been applied in a number of different ways to problems in the Hymenoptera and has proved to have an important and useful set of techniques for answering interesting questions. It is particularly useful for species identification and classification. The more traditional approaches appear to be as sensitive as geometric morphometrics for many problems. A powerful approach is to combine morphometric genetic methods, particularly to help answer questions of systematic and taxonomy.

6. Acknowledgements

Funding for some of this research was provided by a Mount Royal University Internal Research grant. I thank Daniel Lee-Owe for measuring the bee wings.

7. References

Adams, D.C., Rohlf, F.J. & Slice, D.E. (2004) Geometric morphometrics: Ten years of progress following the 'revolution'. Italian Journal of Zoology, 71: 5-16.

Alexander, B. (1991) A cladistic analysis of the genus *Apis*. In: *Diversity in the genus* Apis, Smith, D.R. (ed.) pp. 1-28, Westview Press, Boulder, Colorado.

Aytekin A., M., Terzo, M., Rasmont P. & Çağatay, N., (2007) Landmark Based Geometric morphometric analysis of wing shape in *Sibiricobombus* Vogt (Hymenoptera: Apidae: *Bombus* Latreille). Annales de la Société Entomologique de France, 43: 95-102.

Aytekin,A. M., Rasmont, P. & Çagatay, N. (2003) Molecular and morphometric variation in Bombus terrestris Lucoformis Krüger and Bombus terrestris Dalmatinus Dalla Torre (Hymenoptera: Apidae) Mellifera 3:34-40

Baylac, M., Villemant, C. & Simbolott, G. (2003) Combining geometric morphometrics with pattern recognition for the investigation of species complexes. Biological Journal of the Linnean Society 80: 89-98.

Bertsch, A., Hrabe de Angelis, M. & Przemeck, G.K.H. (2010) A phylogenetic framework for the North American species of the subgenus *Bombus* sensu stricto (*Bombus affinis, B. franklini, B. moderatus, B. occidentalis* & *B. terricola*) based on mitochondrial DNA markers. Beitrage zur Entomologie, 60, 229-242.

Cameron, S.A., Lozier, J.D., Strange, J.P., Koch , J.B., Cordes, N., Solter, L.F. & Griswold, T.L. (2011) Patterns of widespread decline in North American bumble bees. Published online before print January 3, 2011, doi: 10.1073/pnas.1014743108, *PNAS, U.S.A. January 3, 2011*

Clarke, G.M. (1997) The genetic basis of developmental stability. III. Haplo-diploidy: Are males more unstable than females? Evolution 51: 2012-2028.

Colla, S.R. & Packer, L. (2008) Evidence for decline in eastern North American bumblebees (Hymenoptera:Apidae), with special focus on *Bombus affinis*. Biodiv and Cons 17: 1379-1391

Crespi, B.J. & Vanderkist, B.A. (1997) Fluctuating asymmetry in vestigial and functional traits of a haplodiploid insect. Heredity, 79: 624-630.

Cresson (Hymenoptera: Apidae). Canadian Journal of Zoology, 66: 1172-1178.

Danaraddi, C.S. & Viraktamath, S. (2009) Morphometric studies on the stingless bee, *Trigona iridipennis* Smith. Karnataka Journal of Agricultural Science, 22: 796-797.

Eickwort, K.R. (1969) Differential variation of males and females in *Polistes exclamans*. Evolution, 23: 391-405.

Evans, E., Thorpe, R., Jepson, S. & Black, S.H. (2008) Status review of three formerly common species of bumble bee in the subgenus *Bombus*. 63 pp. The Xerces Society, Portland, Oregon.

Fisher, R.A. (1936) The use of multiple measurements in taxonomic problems. Annals of Eugenics, 7: 179-188.

Franklin, H.J. (1913) The Bombidae of the New World. Transactions of the American Entomological Society 38:177-486.

Francoy, T.M., Wittmann, D., Steinhage, V., Drauschke, M., Müller, S., Cunha, D.R., Nascimento, A.M., Figueiredo, V.L.C. Simões ,Z.L.P., De Jong, D., Arias,M.C. & Gonçalves, L.S. (2009) Morphometric and genetic changes in a population of *Apis mellifera* after 34 years of Africanization. Genetics and Molecular Research, 8: 709-717.

Fruciano, C., Tigano, C. & Ferrito, V. (2011) Traditional and geometric morphometrics detect morphological variation of lower pharyngeal jaw in *Coris julis* (Teleostei, Labridae). Italian Journal of Zoology, 78: 320-327.

Gelin, L.F.F., Da Cruz, J.D., Noll, F.B., Giiannotti, E., Dem Santos, G.M., & Bichara-Filho, C.C. (2008) Morphological Caste Studies In The Neotropical Swarm-Founding

Polistinae Wasp *Angiopolybia pallens* (Lepeletier) (Hymenoptera: Vespidae) *Neotropical Entomology 37: 691-701*

Goulson, D., Lye, & B. Darvill, B. (2008) Decline and conservation of bumble bees. Ann. Rev. Ent. 53:191–208

Grixti, J.C., Wonga, L.T., Cameron, S.A. & Favreta, C. (2009) Decline of bumble bees (*Bombus*) in the North American Midwest. Biol. Cons. 142: 75-84

Harder, L.D. (1982) Measurement and estimation of functional proboscis length in bumblebees. Canadian Journal of Zoology, 60: 1073-1079.

Harder, L.D. (1985) Morphology as a predictor of flower choice by bumble bees. Ecology 66: 198-210.

Hastings, J. M., C.W. Holliday and J. R. Coelho (2008) Body size relationship between *Sphecius speciosus* (Hymenoptera: Crabronidae), and their prey: Prey size determines wasp size. Florida Entomologist 91: 657-663.

Hintze, J.L. (1996) *NCSS 6.0.21-2 Statistical System for Windows. User's Manual II.* Hintze, Kaysville, Utah.

Huber, J.T. (2009) Biodiversity of Hymenoptera. In: Foottit, R.G. and Adler, P.H. (eds.), pp. 303-323, *Insect Biodiversity; Science and Society*, Wiley-Blackwell, Chichester, UK.

Huxley, J. (1972) *Problems of Relative Growth 2nd edition,* Dover Publications Inc., New York.

Hastings, J. M., C.W. Holliday and J. R. Coelho. (2010) Size-specific provisioning by cicada killers, *Sphecius speciosus,* (Hymenoptera : Crabronidae) in north Florida. Florida Entomologist 93: 412-421.

Klingenberg, C.P., Badyaev, A.V., Sowry, S.M. & Beckwith, N.J. (2001) Inferring developmental modularity from morphological integration: Analysis of individual variation and asymmetry in bumblebee wings. The American Naturalist, 157: 11-23.

Kozmus, P., Virant-Doberlet, M., Meglič & Dovč, P. (2011) Identification of *Bombus* species based on wing venation structure. Apidologie 42: 472-480.

Marcus, J.M. (2001) The development and evolution of crossveins in insect wings. Journal of Anatomy, 199: 211-216.

Medler, J.T. (1962) Morphometric studies on bumble bees. Annals of the Entomological Society of America, 55: 212-218.

Michener, C.D. (2000) *The Bees of the World.* Johns Hopkins University Press, Baltimore and London.

Milankov, V., Francuski, L., Ludoški, J., Stăhls, G. & Vujié, A. (2010) Estimating genetic and phenotypic diversity in a northern hoverfly reveals lack of heterozygosity correlated with significant fluctuating asymmetry of wing traits. Journal of Insect Conservation, 14: 77-88.

Milliron, H.E. (1971) A monograph of the western hemisphere bumblebees (Hymenoptera: Apidae; Bombinae) - The Genera Bombus, Megabombus Subgenus Bombias. *Memoirs of the Entomological Society of Canada,* 82, 1-80.

O'Neill, K.M., ,Pearce, A.M., O'Neill, R.P, & Miller, R.S. (2010) Offspring Size and Sex Ratio Variation in a Feral Population of Alfalfa Leafcutting Bees (Hymenoptera: Megachilidae) Annals of the Entomological Society of America, 103: 775- 784

Owen, R.E. & McCorquodale, D.B. (1994) Quantitive variation and heritability of postdiapause development time and body size in the Alfalfa leafcutting bee (Hymenoptera: Megachilidae). Annals of the Entomological Society of America, 87:922-927.

Owen, R.E. & Plowright, R.C. (1980) Abdominal pile color dimorphism in the bumble bee *Bombus melanopygus*. Journal of Heredity, 71: 241-247.

Owen, R.E. & Plowright, R.C. (1988) Inheritance of metasomal pile colour variation in the bumble bee *Bombus rufocinctus*

Owen, R.E. (1986) Gene frequency clines at X-linked or haplodiploid loci. *Heredity*, 57:209-219.

Owen, R.E. (1989) Differential size variation of male and female bumble bees (Hymenoptera, Apidae, *Bombus*). Journal of Heredity, 80:39-43.

Owen, R.E. 1988 Body size variation and optimal body size of bumble bee queens (Hymenoptera: Apidae). Canadian Entomologist, 120:19-27.

Owen, R.E., T. L. Whidden & Plowright, R.C. (2010) Genetic and morphometric evidence for the conspecific status of the bumble bees *Bombus melanopygus* and *B. edwardsii* (Hymenoptera, Apidae). Journal of Insect Science, 10:109, available online: insectscience.org/10.109.

Palmer, A. R. (1994) Fluctuating asymmetry analyses: A primer. In: T. A. Markow (ed.) pp. 335-364, *Developmental Instability: Its Origins and Evolutionary Implications*. Kluwer, Dordrecht, Netherlands.

Palmer, A. R., & Strobeck, C. (1986) Fluctuating asymmetry: measurement, analysis, patterns. Annual Review of Ecology and Systematics, 17: 391-421.

Parsons, P.A. (1990) Fluctuating asymmetry: an epigenetic measure of stress. Biological Reviews of the Cambridge Philosophical Society, 65: 131-145.

Pekkarinen, A. (1979) Morphometric, colour and enzyme variation in bumblebees (Hymenoptera, Apidae, *Bombus*) in Fennoscandia and Denmark. Acta Zoologica Fennica 158: 1-60.

Plowright, R.C. & Owen, R.E. (1980) The evolutionary significance of bumble bee color patterns: A mimetic interpretation. Evolution, 34, 622-637.

Plowright, R.C. & Pallett, M.J. (1979) A morphometric study of the taxonomic status of *Bombus sandersoni* (Hymenoptera: Apidae) Canadian Entomologist, 110: 647-654.

Plowright, R.C., & Stephen, W.P. (1973) A numerical taxonomic analysis of the evolutionary relationships of *Bombus* and *Psithyrus* (Apidae: Hymenoptera). Canadian Entomologist, 105: 733-743.

Plowright, R.C., & Stephen, W.P. (1980) The taxonomic status of *Bombus franklini* (Hymenoptera: Apidae). Canadian Entomologist, 112: 475-479.

Pretorius, E. (2005) Using geometric morphometrics to investigate wing dimorphism in males and females of Hymenoptera - a case study based on the genus *Tachysphex* Kohl (Hymenoptera: Specidae: Larinae). Australian Journal of Entomology, 44: 113-121.

Prŷs-Jones, O.E. & Corbet, S.A. (1987) *Bumblebees*. Cambridge University Press, Cambridge.

Richards, K.W. (1984) Alfalfa leafcutter bee management in Western Canada. Agriculture Canada Publication 1495E: 1-53.

Rohlf F. J. & Marcus L. F. (1993) A revolution in morphometrics. Trends Ecol. Evol., 8: 129-132

Rohlf, F.J. & Slice, D. (1990) Extensions of the procrustes method for the optimal superimposition of landmarks, Systematic Zoology, 39: 40-59.

Rosenberg, M.S. (1997) Evolution of shape: differences between the major and minor chelipeds of *Uca pugnax* (Decapoda: Ocypodidae). Journal of Crustacean Biology, 17: 52-59.

Rothschild, M. (1979) Factors influencing size and sex ratio in *Megachile rotundata* (Hymenoptera, Megachilide). Journal of the Kansas Entomological Society, 52: 392-401.

Ruttner, F. (1986) Geographical variability and classification. In: *Bee Genetics and Breeding*, Rinderer, T.E. (ed.) pp. 23-56 Academic Press, inc., New York.

Scholl, A., E. Obrecht & Owen, R.E. (1990) The genetic relationship between *Bombus moderatus* Cresson and the *Bombus lucorum* Auct. species complex. Canadian Journal of Zoology, 68, 2264-2268.

Scholl, A., Thorp, R.W., Owen, R.E. & Obrecht, E. (1992) Specific distinctiveness of *Bombus nevadensis* Cresson and *B.auricomus* (Robertson)(Hymenoptera: Apidae) - enzyme electrophoretic data. Journal of the Kansas Entomological Society, 65, 134-140.

Sharkey, M.J. & Roy, A. (2002) Phylogeny of the Hymenoptera: a reanalysis of the Ronquist *et al.* (1999) reanalysis, with an emphasis on wing venation and apocritan relationships. Zoologica Scripta, 31, 57–66

Sharkey, M.J. (2007) Phylogeny and classification of Hymenoptera. Zootaxa, 1668, 521–548

Shih, C., Feng, H., Liu, C., Zhao, Y., & Ren, D. (2010) Morphology, phylogeny, evolution, and dispersal of pelecinid wasps (Hymenoptera: Pelecinidae) over 165 million years. Annals of the Entomological Society of America, 103: 875-885.

Silva, M.C., Lomônaco, C., Augusto, S.C. & Kerr, W.E. (2009) Climatic and anthropic influence on size and fluctuating asymmetry of Euglossine bees (Hymenoptera, Apidae) in a semideciduous seasonal forest reserve. Genetics and Molecular Research, 8: 730-737.

Slice, D.E., Bookstein, F.L., Marcus, L. F. & Rohlf, F.J. (Revised Feb. 12, 2009) A Glossary for Geometric Morphometrics In: *Morphometrics at SUNY Stony Broo,k* Date of access, 30.09.2011, Available from: http://life.bio.sunysb.edu/morph/

Smith, D.R., Crespi, B.J. & Bookstein, F.L. (1997) Fluctuating asymmetry in the honey bee, *Apis mellifera*: effects of ploidy and hybridization. Journal of Evolutionary Biology, 10: 551-574.

Sneath, P.H.A. & Sokal, R.R. (1973) *Numerical Taxonomy*. W.H. Freeman & Co., San Franscisco.

Stephen, W.P. (1957) Bumble Bees of Western America - (Hymenoptera: Apoidea). *Technical Bulletin*, 40, 2-163.

Thorp, R.W., Horning, D.S., and Dunning, L.L. (1983) Bumble bees and cuckoo bumblebees of California (Hymenoptera: Apidae). *Bulletin of the California Insect Survey* 23:1-79.

Tofilski, A. (2004) DrawWing, aprogram for numerical description of insect wings. 5pp. *Journal of Insect Science*, 4:17, Available online: http://www.insectscience.org/4.17

Tüzün, A. (2009) Significance of wing morphometry in distinguishing some of the hymenoptera species. African Journal of Biotechnology 8: 3353-3363.

Whidden, T.L. (2002) Applications of randomly amplified and microsatellite DNA to problems in bumble bee biology. 155 pp. Doctoral dissertation, University of Calgary.

Williams, P.H. (1998) An annotated checklist of bumble bees with an analysis of patterns of description (Hymenoptea: Apidae, Bombini). *Bulletin of the National History Museum of London* 67, 79-152.

Wilson, E.O. (1971) *The Insect Societies*. The Belknap Press of Harvard University Press, Cambridge, Mass.

2

The Mosquito Fauna: From Metric Disparity to Species Diversity

Jean-Pierre Dujardin[1], P. Thongsripong[2] and Amy B. Henry[3]
[1]*UMR 5090 MIVEGEC, Avenue Agropolis, IRD, Montpellier*
[2]*Department of Tropical Medicine, Medical Microbiology, and Pharmacology, University of Hawaii at Manoa, Honolulu, Hawaii*
[3]*Department of Microbiology, University of Hawaii at Manoa, Honolulu, Hawaii*
[1]*France*
[2,3]*USA*

1. Introduction

Biodiversity, encompassing diversity of genes, species, and ecosystems, is fundamental to biology (Gaston & Spicer, 2004), yet tools to monitor it are insufficient. Biodiversity can be estimated by using the number of species (species richness) in a community, and/or by this number together with the proportion of species (species evenness), and/or by other more indirect estimators among which is morphological variation.

Studies on morphological and biological diversity have highlighted the complexity of the diversity structure, that is, the relationship between morphological and taxonomic diversity (Foote, 1992). Most studies acknowledge a certain level of dissociation between morphological diversity and species richness, suggesting that taxonomic and morphological diversity patterns are distinct ones (Foote, 1993; Moyne & Neige, 2007; Roy et al., 2001; Roy & Foote, 1997; Vasil'ev et al., 2010). Occasionally, the use of metric diversity as a proxy for species richness has been suggested (Dolan et al., 2006).

However, appropriate morphological disparity metrics and sets of morphological characters are not clearly defined (Navarro, 2003; Roy & Foote, 1997; Wills et al., 1994). Although most studies have used one or two measures of disparity to quantify and characterize the occupation of morphospace, multiple measures might be necessary to fully detect changes in patterns of morphospace occupation (Ciampaglio et al., 2001).

Moreover, organisms present an indefinitely large number of potentially quantifiable traits, and in practice only a small number of features can be studied. Therefore, we cannot strictly measure morphological diversity, but diversity with respect to some set of traits. The common practice is to seek broad coverage of morphology, using as many characters as possible. The present study is disputing this common practice, showing that the opposite strategy could be more informative: to seek for elementary coverage of morphology.

Our hypothesis is that when grouping many traits morphological variation becomes too complex to validly reflect a single factor like species richness. Organismal morphology is the result of many biological causes that are not only evolutionary factors but also environmental and historical. Instead of considering a conglomerate of traits to " cover " the organismal morphology, we suggest to decompose the global morphology into more elementary units and test their variation against species richness.

A set of anatomical landmarks in the wing represents a suitable tool to explore that hypothesis since it can be decomposed into subsets of different landmarks. In our approach, the complete set of anatomical landmarks of the wing would represent the broadest morphological coverage, and its decomposition into smaller configurations of landmarks would provide more elementary units of morphology. Do different subsets (combinations) of landmarks reflect biodiversity in the same way as the total set? Landmark-based geometric morphometrics, which is applied here, provides a convenient way for measuring morphological variety, requiring only the recognition of homologous landmarks in all individuals under comparison. This condition applies very well to mosquito wings, because their venation pattern is almost identical among different species and higher taxa, including different tribes.

2. Materials & methods

2.1 The insects

We used a total sample of 480 individuals (one wing per individual). They were tentatively identified using available morphological keys (Rattanarithikul et al., 2005; Rattanarithikul, Harrison, Panthusiri, Peyton & Coleman, 2006; Rattanarithikul, Harrison, Harbach, Panthusiri & Coleman, 2006.; Rattanarithikul et al., 2010). A total of 10 genera and 43 species of Culicidae (mosquitoes) was found, with unidentified species pooled into one putative "species" (Table 1). This collection was a representative set of higher taxa within Culicidae: it contained indeed the two subfamilies (Anophelinae and Culicinae) and, within the Culicinae, 6 tribes out of 11.

2.2 Mosquito collection

The mosquito collection was done during the rainy season of 2008 (June-August) along a forest-agro-urban landscape gradient within Nakhon Nayok province, central Thailand. Six habitat types: forest, fragmented forest, rice field, rural, suburban, and urban were identified and characterized across the landscape gradient (Table 1). For each habitat type, four sites were picked as representative of similar habitat range. Forest sites were situated along the border of the pristine Khao Yai National Park. Fragmented forest sites were on the edge of disturbed vegetation patch not far from the National Park, where human settlements were sparse and traditional small-scale agricultures were practiced. Rice field sites were further away from the National park and situated in the lowland closer to the main river hence the big-scale and irrigated rice agricultures were possible. The rural, suburban, and urban sites distributed based on the distance from the centre of town. The rural sites were more than 7 km from town; the suburban were within 5 km from town; and the urban sites were in the center of the town.

Four types of adult mosquito trap: BG sentinel, Mosquito Magnet, CDC UV light traps, and CDC backpack aspirator, were used in order to maximize the variety of mosquito samples collected. In each site, mosquitoes were collected for 24 hours using 8 BG traps, 2 Mosquito Magnet traps, and 3 of 3 to 10 minute-long aspirations for the day trapping, and 8 UV light traps, 8 BG traps, and 2 Mosquito Magnet traps for the night trapping. A total of over

80,000 mosquitoes were collected. For the morphometric study, only a subset of the female mosquitoes (Table 1) were examined.

Genus	Species	*n*	F	FF	R	RF	SU	U
Aedes	*aegypti*	*12*	.	.	1	3	5	3
	albopictus	*7*	.	4	2	1	.	.
	lynetopennis	*14*	.	4	10	.	.	.
	mediolineatus	*12*	.	6	6	.	.	.
	vexans	*25*	.	7	16	.	2	.
	unknown	*20*	1	5	7	2	5	.
Aedomyia	*catasticta*	*1*	.	1	.	.	.	.
Anopheles	*baezai*	*1*	.	.	.	.	1	.
	barbirostris	*10*	2	3	.	1	3	1
	kochi	*6*	.	6	.	.	.	.
	minimus	*2*	.	2	.	.	.	.
	peditaeniatus	*5*	.	.	1	2	2	.
	phillippines	*5*	.	3	2	.	.	.
	tessellatus	*9*	.	4	.	.	5	.
	vagus	*37*	.	21	12	.	1	3
	unknown	*7*	.	7	.	.	.	.
Armigeres	*dentatus*	*1*	1	.	.	.	.	.
	magnus	*2*	2	.	.	.	.	.
	malayi	*1*	.	1	.	.	.	.
	subalbatus	*10*	2	4	1	.	1	2
	unknown	*2*	.	2	.	.	.	.
Coquillettidia	*crassipes*	*8*	.	.	.	8	.	.
	unknown	*5*	.	5	.	.	.	.
Culex	*bitaeniorhynchus*	*22*	5	4	1	4	7	1
	brevipalpis	*7*	.	2	2	2	.	1
	fuscocephala	*6*	.	5	.	.	.	1
	gelidus	*18*	.	4	1	5	4	4
	mocthogenes	*1*	.	1	.	.	.	.
	nigropunctatus	*12*	10	.	.	.	1	1
	quinquefasciatus	*11*	.	1	.	.	4	6
	sinensis	*28*	3	11	1	7	6	.
	tritaeniorhynchus	*1*	.	1	.	.	.	.
	vishnui	*53*	1	29	8	1	11	3
	unknown	*24*	2	14	1	3	2	2
Ficalbia	*minima*	*3*	.		3		.	.
Heizmania	unknown	*7*	7		.		.	.
Mansonia	*annulifera*	*5*	.	3	.		.	2
	indiana	*6*	.		.	1	5	.
	uniformis	*11*	.	7	.	1	3	.
	chamberlainai	*5*	.		3	1	1	.
	hybrida	*7*	2	2	2	1	.	.
	luzonensis	*7*	2	1	1	.	.	3
	metallica	2	.	.	.	.	2	.
	unknown	*1*	.	1	.		.	.

Genus	Species	*n*	F	FF	R	RF	SU	U
Uranotaenia	*campestris*	2	.	2	.	.	.	.
	lateralis	1	.	.	.	.	1	.
	lutescens	4	.	.	.	1	3	.
	micans	7	.	1	.	3	3	.
	nivipleuraura	3	2	.	.	.	1	.
	subnormalis	2	.	.	.	.	2	.
	unknown	22	17	2	1	1		1

Table 1. List of genera and species of Culicidae identified on morphological ground, with their repartition according to the habitat. F, forest; FF, fragmented forest; RF, rice field; R, rural; SU, semi-urban and U, urban. For statistical tests, unknown species have been pooled into one single taxon (92 specimens). *n*, number of specimens submitted to morphometric analyses.

2.3 Shape of the wing

The shape of the mosquito wings was described by 13 landmarks (see Fig. 1).

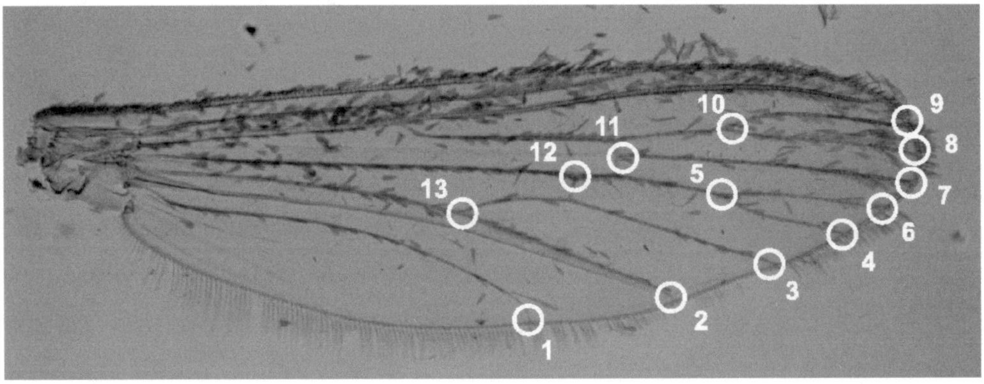

Fig. 1. Mosquito wing. Landmarks are labelled according to the order of digitization.

The decomposition of shape used a total of 254 different LM configurations out of the 7814 possible ones, i.e. the totality (13) of landmarks (LM), the 13 combinations of 12 LM, and 240 combinations of decreasing numbers of LM. For each subset of LM, going from 11 LM to 4 LM, 30 different combinations were tested (see Table 2).

The above testing of 254 LM configurations was replicated on ten different species sequences (see Table 3).

2.4 Species sequence

A sequence refers here to a succession of species assemblages with increasing richness. The mosquitoes were randomly sampled with replacement into 22 assemblages of increasing species richness. These 22 assemblages constituted a sequence where each unit represented a community with different species richness, starting from 2 species, 4 species, 6 species, and so on until 44 species. A partial representation of a sequence is shown Table 3; ten such sequences were constituted. The range of categorical units (44 species) was not modified.

	(4)	(5)	0	(10)	(11)
1	8-9-11-13	1-4-6-8-11	0	1-2-5-7-8-10-11-12-13	1-3-4-5-7-8-10-11-12-13
2	4-5-7-13	3-5-9-10-11	0	1-3-4-5-6-7-9-10-11-12	1-2-3-4-5-6-9-10-11-12-13
3	2-6-7-13	5-9-10-11-12	0	1-2-3-4-5-7-8-10-11-12	1-2-3-4-7-8-9-10-11-12-13
4	2-3-8-13	4-6-7-10-13	0	2-3-5-6-7-8-9-10-12-13	2-3-4-5-7-8-9-10-11-12-13
5	1-2-3-10	1-4-6-9-13	0	3-4-5-6-7-8-9-10-11-13	1-2-3-6-7-8-9-10-11-12-13
6	4-5-12-13	1-2-3-4-5	0	3-4-5-6-8-9-10-11-12-13	1-2-4-6-7-8-9-10-11-12-13
7	5-8-10-13	2-3-7-9-11	0	1-2-4-6-7-8-9-10-11-12	2-3-4-5-6-7-8-10-11-12-13
8	2-9-12-13	2-3-5-7-12	0	1-2-6-7-8-9-10-11-12-13	1-2-3-4-5-6-7-9-11-12-13
9	1-3-8-12	1-3-7-8-10	0	1-2-3-5-6-7-9-11-12-13	2-3-4-6-7-8-9-10-11-12-13
10	1-4-5-9	1-5-7-12-13	0	1-2-4-5-6-9-10-11-12-13	1-2-3-4-5-6-7-9-10-11-12
11	6-8-10-13	2-4-7-8-10	0	2-3-4-6-7-8-9-10-11-13	1-2-3-5-6-7-8-9-10-11-12
12	4-7-8-9	3-4-8-11-12	0	1-2-3-4-5-6-7-8-12-13	1-3-4-5-6-7-8-9-10-11-12
13	1-3-6-9	1-5-8-9-10	0	1-2-3-5-7-8-10-11-12-13	1-2-3-4-5-7-8-9-10-11-13
14	1-5-8-10	3-5-9-11-12	0	2-3-4-5-6-7-8-9-10-13	1-2-4-5-6-7-8-9-10-12-13
15	1-4-8-9	3-8-9-10-11	0	1-2-4-5-6-7-8-10-12-13	1-2-3-4-5-6-7-8-9-10-12
0	0	.	0	0	0
30	3-5-8-12	2-5-8-9-13	0	1-2-3-4-5-7-9-11-12-13	1-2-3-4-5-6-7-8-9-12-13

Table 2. A partial representation of the sets of landmarks used to examine a sequence of increasing taxonomic richness (see Table 3). Landmarks were numbered from 1 to 13 (Fig. 1). The first row refers to the number of landmarks (between brackets). The first column enumerates the 30 combinations of landmarks randomly generated for each number of landmarks. The second column partially shows the 30 combinations to represent 4 landmarks. To save space, only some combinations for 4, 5, 10 and 11 landmarks are represented here. To these 240 (30*8) combinations, we added the 13 combinations of 12 landmarks, and the total number of landmarks (13).

2.5 Species assemblage

In the building of sample sets of increasing taxonomic richness (a sequence), species were not " added " to previous ones, they were resampled at each step from the total available. This can be seen in the sequence shown Table 3. The program was not simulating a temporal variation where species progressively accumulate in a given environment, but a spatial sampling of taxa where groups represent communities of various taxonomic richness.

2.6 Sampling individuals

To measure metric disparity (MD), a total of 50 specimens by assemblage was typically used. However, for assemblages of low SR (2 to 8 species), the sample size could be less than 50 (28.6 +- 9.6). On average, the sample size was 45.8 +- 8.7. The abundance of each species within an assemblage was approximately the same. For instance, an assemblage of 10 species contained approximately 5 specimens per species, while an assemblage of 24 species could contain for instance 20 species with 2 individuals and 4 species containing 3 individuals. This situation of high evenness is generally not the one found in natural conditions.

2.7 Metric disparity

To estimate morphological diversity, we considered only the geometric, landmark-based approach. Morphological diversity was estimated by the metric disparity (MD) index. For

(SR)		*(2)*		*(4)*		*(6)*		*(8)*	*(etc.)*	*(40)*		*(44)*
1	13	21										
etc.	.	.										
30	3	14										
1	15	18	23	34								
etc.	.	.	.	.								
30	3	12	14	38								
1	7	17	22	35	37	41						
etc.	.	.	.	.	.	.						
30	3	5	6	10	26	38						
1	11	12	16	22	24	35	38	42				
etc.	.	.	.	.	.	.	.	.				
30	3	6	7	13	18	20	26	37				
1	6	9	18	19	22	28	34	38	etc.			
etc.	.	.	.	.	.	.	.	.	etc.			
30	3	6	9	13	15	24	27	38	etc.			
etc.	etc.	etc.	etc.	etc.	etc.	etc.	etc.	etc.	etc.			
1	1	2	5	6	8	11	12	13	etc.	41		
etc.	.	.	.	.	.	.	.	.	etc.	.		
30	1	3	4	6	7	8	10	11	etc.	44		
1-30	1	2	3	4	5	6	7	8	etc.	42	43	44

Table 3. A partial representation of a sequence of increasing taxonomic richness (numbers between brackets) as used by our simulation programme. Each number represents a mosquito species. There were 44 different taxa. The first set of rows shows two assemblages of two randomly selected species (species "13", species "21" and species "3", species "14") among the 30 species pairs randomly generated. The second set of rows shows two such assemblages of 4 species among the 30 random combinations of 4 species, and so on till reaching an assemblage of 44 species. This makes a total of 22 assemblages of increasing species richness, each one sampled 30 times among species. The last assemblage, containing the totality of the species, was sampled 30 times among individuals (*1-30*). Ten such sequences were generated, and each one was explored by using 254 different landmarks configurations (Table 2).

each of N wings, after Procrustes superposition using the Generalized Procrustes Algorithm (GPA) (Rohlf, 1990) and partial warps (PW) computation as in Rohlf (1996), the sum of squared PW was obtained. This sum was divided by the degrees of freedom (N-1) to compute MD (Zelditch et al., 2004). To estimate MD for an assemblage with, for instance, 2 species, 30 random pairs of species were used (see Table 3) and the average MD value was considered.

2.8 Habitat heterogeneity

The sample composition did not allow valid statistics using the habitat as a categorical unit (instead of the species). To evaluate the importance of the environment on the metric disparity (MD), we performed a simple two-way ANOVA with taxa and habitat as effects. The variable used was the individual sum of squared PW because this sum is the term directly used to compute MD (see above). Five ANOVA were performed, one using the totality of LM to compute the PW, and four using selected combinations of LM. The latter were chosen according to their relationship to species diversity: they were two configurations of LM that produced MD highly correlated to the species richness (SR), and two others not related to SR (see Table 7).

2.9 Biodiversity

To estimate the biodiversity, we used the species richness index (SR, or the number of categorical units) which is the total number of species found in the community (an assemblage). Since evenness was maintained as high as possible, biodiversity indexes like the Shannon index (Shannon & Weaver, 1949) or the Simpson one (Simpson, 1949) were not examined.

2.10 Diversity structure

Diversity structure (DS) is described here as the relationship between MD and SR (Foote, 1992). It was estimated by the determination coefficients (squared linear correlation coefficients) (see Table 5), and illustrated graphically (Fig. 2). An average estimate of MD was used during all correlations. For each specific combination of LM, at each level of SR, MD was an average value derived from the 30 different assemblages (Table 3). For one specific number of LM, for example 4, this average was performed taking into account also 30 combinations of 4 LM (Table 2).

2.11 Software

A special TclTk (`http://www.tcl.tk/software/tcltk/`) script was written where Procrustes superposition (GPA), partial warps (PW) computations, as well as metric diversity (MD) estimations, made use of procedures extracted from the CLIC package (`http://www.mpl.ird.fr/morphometrics/clic/index.html`). Table 3 was computed using STATA (?). Figure 2 used the GNUMERIC spreadsheet (`http://projects.gnome.org/gnumeric/`).

3. Results

Globally, the effects of SR on various LM configurations of the same wing confirmed our initial hypothesis: different aspects of shape did not vary the same way in response to the same factor. The relationship between species richness (SR) and metric disparity (MD) was estimated by the determination coefficient (see DS, for "diversity structure" in Tables 4, 5 and 6). This coefficient does not inform about the sign of the correlation between SR and MD, which was not necessary since these correlation coefficients were all positive. The species richness contributed to the metric disparity according to the number of landmarks used to compute MD (Tables 4 and 5), and also according to the identity of landmarks involved (Table 6). We subdivide our results according to these two aspects of shape composition: the number (see paragraph 3.1) and the identity (see paragraph 3.2) of LM. We then show the ANOVA output for some remarquable configurations of LM (high DS, low DS): it estimates the possible role of habitat heterogeneity on the metric disparity (see paragraph 3.3).

3.1 Species richness and the number of landmarks

We observed that the configurations involving a low number of landmarks varied more frequently in very close accordance with species richness. Actually, for the same species arrangements, some configurations of the wing involving a very low number of landmarks (LNLM configurations) had either very high (column "DS > 75", see Table 4) or very low (column "DS < 50", see Table 4) prediction power on species diversity, while more complex anatomical configurations of the same wing showed a more stable but less predictive relationship with SR.

LM_nb	total	sampled	DS $\leq$ 55	55 $<$ DS $<$ 75	DS $\geq$ 75
4	715	(4%)	26%	45%	29%
5	1287	(2%)	21%	71%	8%
6	1716	(2%)	17%	71%	12%
7	1716	(2%)	12%	82%	9%
8	1287	(2%)	16%	77%	7%
9	715	(4%)	6%	87%	7%
10	286	(10%)	10%	83%	6%
11	78	(38%)	5%	86%	9%
12	13	(100%)	0%	98%	2%
13	1	(100%)	0%	100%	0%

Table 4. For each number of landmarks (LM_nb), the column "total" gives the total number of possible configurations among the possible landmark positions on the wing (a total of 13), the column "sampled" indicates the percentage of such configurations that have been studied; except for 12 and 13 LM, we always examined 30 random configurations for each number of landmark. Thus, 4% for instance is 30 out of 715. DS, or the "diversity structure" was estimated here by the determination coefficient (expressed in percentages) between metric disparity and species richness. The three last columns refers to the frequency at which a given DS was observed. It can be seen that a determination coefficient lower than 55% or higher than 75% between species richness and metric disparity was observed more frequently when using a low number of LM

Both the best and worst DS scores between SR and MD were obtained with configurations made from a low number of landmarks (LNLM). The range of scores progressively decreased with the addition of more LM (see column SD of Table 5). With more numerous LM however, the best predictive values did not reach such high levels as with fewer LM. Thus, more LM meant a more stable assessment of diversity structure (DS, or the relationship between MD and species richness), with no occurrence of very high values (Table 5).

3.2 Species richness and the identity of landmarks

Not only were we able to disclose different diversity structures according to the number of landmarks, but also according to specific configurations of landmarks. A partial output is presented Table 6. The highest determination coefficient (94%) observed was obtained with a specific combination of 4 LM (see 2-3-8-13, Table 4); other combinations involving the same number of LM gave much lower predictability (see 1-3-8-12, Table 4). It could be as low as 27% (see the columns MAX and MIN of Table 5).

NLM	DS	SD	MIN	MAX	STABILITY
4	64.7%	13.8	27	94	5.8
5	63.4%	7.7	35	85	6.0
6	65.4%	8.3	40	89	5.7
7	65.3%	5.8	41	86	6.4
8	64.1%	6.5	40	83	6.6
9	66.4%	3.6	49	83	6.1
10	64.5%	3.9	47	79	6.8
11	65.8%	3.4	48	80	6.3
12	67.5%	2.5	59	71	2.2
13	64.0%	0.6	57	62	3.5

Table 5. Determination coefficients (column DS) as percentages representing the contribution of species richness (SR) to metric disparity (MD) according to the number of landmarks (NLM) used to compute MD. These coefficients allow comparisons of the diversity structures (DS), i.e. the relationship between MD and SR according to NLM. Columns DS lists the determination coefficients and SD their standard deviation; MIN is the minimum value of DS and MAX its maximum. The table, except last column, represents an average DS derived from an average sequence of groups having increased SR; it was computed from the ten replicated sequences used in the study (see Table 3). The last column (STABILITY) is the standard deviation of the DS mean scores obtained from one sequence to another; it indicates how stable were the DS for the same LM configurations across 10 different random sequences of species. For each number of landmarks (each row), DS and SD values were averaged over 30 combinations of different landmarks (except for 12 LM which had only 13 possible configurations and of course for 13LM, see Table 2) and over ten replicated sequences, each one providing an average estimate from 30 random assemblages of species (Table 3)

For the same specimens and the same species arrangements, some shape components (i.e. landmarks configurations) of the wing varied in accordance with the number of species and others did not. Furthermore, a LM configuration highly predictive of SR, like the set of landmarks 2, 3, 8, 13 or the landmarks 2, 6, 7, 13, remained predictive regardless of the species sequences. The same observation applied for non-predictive sets of landmarks, like for instance the set of landmarks 1, 4, 5, 6 or 1, 3, 6, 9. This stability was verified across the ten replicates (see last column of Table 5).

3.3 The habitat heterogeneity

The highly predictive LNLM configurations, like 2-3-8-13 and 2-6-7-13 used in the ANOVA, were both affected by species richness only ($P < 0.0001$), not by the habitat heterogeneity ($P > 0.0500$), while the TNLM and two poorly predictive LNLM (1-4-5-6 and 1-3-6-9) were affected by both species and habitat (Table 7).

4. Discussion

To explore the diversity structure (DS), i.e. the relationship between metric and biological diversity, our model tested the effect of species richness (SR) on the metric disparity (MD) computed from 254 possible combinations of landmarks. Our data showed that the DS

	(4)	MIN	MAX	(11)	MIN	MAX
1	8-9-11-13	71	93	1-3-4-5-7-8-10-11-12-13	56	75
2	4-5-7-13	44	62	1-2-3-4-5-6-9-10-11-12-13	55	67
3	2-6-7-13	71	90	1-2-3-4-7-8-9-10-11-12-13	58	70
4	2-3-8-13	77	94	2-3-4-5-7-8-9-10-11-12-13	58	76
5	1-2-3-10	44	72	1-2-3-6-7-8-9-10-11-12-13	57	76
6	4-5-12-13	54	74	1-2-4-6-7-8-9-10-11-12-13	57	76
7	5-8-10-13	52	72	2-3-4-5-6-7-8-10-11-12-13	58	75
8	2-9-12-13	62	80	1-2-3-4-5-6-7-9-11-12-13	56	70
9	1-3-8-12	37	48	2-3-4-6-7-8-9-10-11-12-13	61	77
10	1-4-5-9	39	62	1-2-3-4-5-6-7-9-10-11-12	49	63
11	6-8-10-13	51	71	1-2-3-5-6-7-8-9-10-11-12	49	68
12	4-7-8-9	74	89	1-3-4-5-6-7-8-9-10-11-12	50	68
13	1-3-6-9	64	78	1-2-3-4-5-7-8-9-10-11-13	56	77
14	1-5-8-10	47	65	1-2-4-5-6-7-8-9-10-12-13	52	73
15	1-4-8-9	65	79	1-2-3-4-5-6-7-8-9-10-12	48	67
.	.	.	.	.	.	.
30	3-5-8-12	42	63	1-2-3-4-5-6-7-8-9-12-13	54	76

Table 6. Minimum (MIN) and maximum (MAX) determination coefficients as percentages representing the contribution of species richness to metric disparity according to the configurations of landmarks used. The 8-9-11-13 formula means the configuration of four landmarks using landmarks 8, 9, 11 and 13 as represented in Fig. 1. As in Table 2 the LM configurations are classified according to the number of LM (number between brackets). To save sapece, we present only a partial output of the data for 4LM and 11LM. For each configuration of landmarks (each row), values were obtained from ten replicated sequences, each one providing an average estimate from 30 random assemblages of species (Table 3). It can be seen that the contribution of species richness to metric disparity (MD) can be very high when MD is computed from a low number of LM (column 4LM), which does not seem to be the case for configurations involving more landmarks (column 11LM). Table 5 provides the average DS over all the tested configurations for each number of LM (not only 4 and 11).

depended on the aspects of shape that were considered (the number of LM, the configuration of LM).

4.1 Our model

The ability to detect morphological trends and occupation patterns within morphospace might depend on using the appropriate measure(s) of disparity. Since there is no clear indication about which index is best, some authors used a variety of measurements (Ciampaglio et al., 2001; Navarro, 2003). Our model presented results regarding only the MD index: it is the most commonly applied measurement of morphological disparity in landmark-based geometric studies (Zelditch et al., 2004).

In trying to include 50 individuals per group within our model, we were unable to consider the effect of sampling variation. However, metric disparity, as measured here, is relatively insensitive to sample size variation and has been shown to be a stable estimate when using a sample of 50 individuals (Navarro, 2003).

Source	Partial SS	df	MS	F	Prob > F
Model (13 LM)	0.036945	48	0.000770	14.710	0.000000
species richness	0.030978	43	0.000720	13.770	0.000000
habitat heterogeneity	0.005266	5	0.001053	20.130	0.000000
Residual	0.022545	431	0.000052		
Total	0.059491	479	0.000124		

Source	Partial SS	df	MS	F	Prob > F
Model (2-3-8-13)	0.001168	48	0.000770	10.240	0.000000
species richness	0.001126	43	0.000024	11.020	0.000000
habitat heterogeneity	0.000025	5	0.000005	2.140	0.059800
Residual	0.001025	431	0.000002		
Total	0.002193	479	0.000005		

Table 7. Two-way ANOVA showing the contribution of species richness and habitat heterogeneity to the individual sum of squared PW: this sum per individual is the value on which metric disparity is directly estimated (as an average). The ANOVA was performed on values computed from the total number of landmarks (TNLM configuration, top) or from the following configuration of four landmarks: 2-3-8-13 (bottom). For the latter (and also for LNLM highly predictive configurations like 2-6-7-13, not shown here), only species richness contributed to metric variation. Both species richness and habitat heterogeneity contributed to the variation of the total set of landmarks (as well as of LNLM poorly predictive configurations like 1-4-6-9 or 1-3-5-6, not shown here)

The model did not randomize the total number of species, which was always fixed to 44 taxa, nor did it randomize the sequence of species richness (2, 4, 6,..., 44), two parameters that are likely to affect the diversity structure of some landmark configurations.

There is however no *a priori* reason to think that these shortcomings would reduce the interest to use smaller configurations of shape to simplify the interpretation of the diversity structure.

More critical may be that our model did not take into account variation in species evenness. In each assemblage of species for a given sequence, the model was designed to get an approximately equal number of specimens within each species. In the natural conditions, there is generally no such evenness, and, because the MD index is an average value, this might affect the correlation between MD and SR. Rare taxa are often assumed to exhibit unusual morphologies because of specialized life habits and could thus contribute disproportionately to the disparity of an assemblage (Deline, 2009). But wings of mosquitoes, because they are generally very similar, are unlikely to have such an effect. Their general similarity may cause the opposite effect to take place: rare taxa may fill the mophospace among the common species and would likely lower MD (Deline, 2009).

Finally, in our simulation the successive assemblages contained an increasing number of species randomly selected from a total pool of mosquitoes, they were not obtained by " adding " species to previously selected ones. Because of this specific design, our simulation is not relevant to temporal follow-up of metric diversity, however it is specific to the spatial comparison of biodiversity.

4.2 Species diversity: The number of landmarks

Although correlations between MD and SR could be weak or not significant, they were positive. A positive correlation has an intuitive explanation: more diversity means more forms. Weak and not significant correlation between SR and MD obtained from some combinations of LM could be the consequence of a few species immediately occupying the extremes of the morphospace. In that situation, by filling in the morphospace between disparate taxa, increasing taxonomic diversity was not able to have a significant effect on MD estimates (Roy & Foote, 1997).

The intensity of the correlation between SR and MD depended on the number of LM. Unexpectedly, the low number of LM (LNLM) configurations gave the best predictive power on SR (Fig. 2, left side). This is counter-intuitive since it is the current belief that more shape would have more taxonomic contents and hence would be more useful to distinguish species. We verified that by using one of the most predictive configuration of 4 LM (2-3-8-13) only 30% of individuals could be correctly assigned to their species, while more than 75% could be correctly attributed with 13 LM (the TNLM configuration). How could it be that a better discriminating configuration of LM could give a lower taxonomic prediction than a poorly discriminating one? The answer could lie in the observation that a LNLM configuration also could have no relationship at all with SR, and that the TNLM configuration is gathering both the predictive and non-predictive subsets of LM. Thus, the mixing of better predictive with poorly predictive components of shape into a larger set of LM was likely to mix opposite trends and blur the relationship between MD and SR (Fig. 2, right side). In other words, when a larger number of LM is used as a proxy for SR, a larger number of influences is also allowed which is not limited to SR. This explanation raises the question whether in our material there were other influences affecting the variation of the LM configurations that were not related to SR.

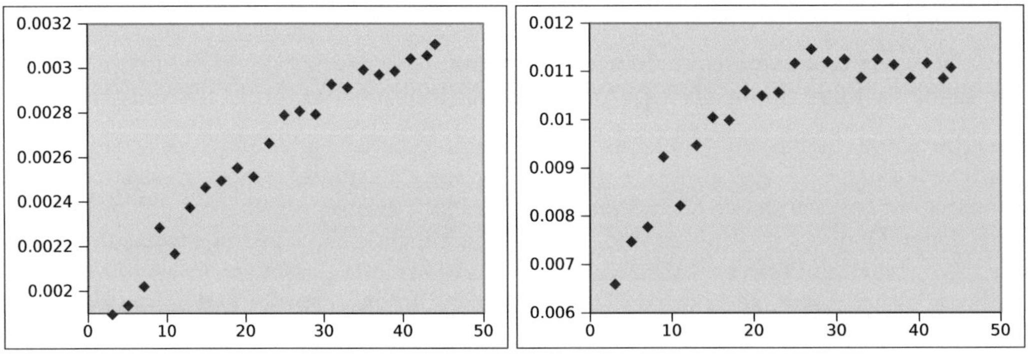

Fig. 2. Two graphs representing the diversity structure (DS), i.e. the relationship between species richness (SR, on the horizontal axis) and metric diversity (MD, on vertical axis). The left graph shows the DS appearing when using a set of four landmarks (namely the landmarks 2, 3, 8 and 13; see Fig. 1). The right graph shows the DS using the total number (13) of LM. When using the totality of LM (right graph) there is a first list of increasing MD values up to a SR of 30 species. This first list is common to the two graphs. In the right one however, after 30 species, there is a " plateau " breaking the global correlation between SR and DM

4.3 Species diversity: The identity of landmarks

For a given number of landmarks, the same configurations produced high (or low) correlation with SR regardless of the taxonomic composition of the species sequence. For instance, the 2-3-8-13 configuration always produced high correlation with SR regardless of the sequence of species considered, and the 1-3-6-9 configuration always produced one of the worst correlations. Thus, there seemed to be specific configurations of LM that could reflect the number of species, and others that could not, regardless of the species sequence used. This observation raises two questions:

- Are there any LM configurations that could be used in mosquitoes as a proxy for species richness estimations?
- What are the sources (other than randomness) that generated variation for the LM configurations that were not influenced by the species richness?

The first question is about the effect of taxonomic diversity on wing shape variation: can the latter be used as a proxy for indirect species richness estimation?

Before starting to answer that question, one could ask what the interests are in making an indirect estimation of species richness. The answer is about time. Metric disparity extracted from wing venation is much faster to obtain than metrics requiring a taxonomic identification of each collected individual. Mosquitoes are often distinguished on the basis of labile characters which may be lost on damaged specimens, making species diagnostic difficult in a group where species are numerous (3500 species) and taxonomists are unfortunately not. An indirect but fast comparative estimation of SR would be welcome.

Our results showed that some landmarks configurations were highly predictive for SR. If these landmark configurations were known in advance, their variation could be used indeed as a proxy for SR. In our study, their performances seemed stable regardless of species arrangement in different groups (Table 5, last column). However, whether the high scoring of these LM configurations would be maintained in other studies on other mosquito species is still to be investigated.

The second question refers to other possible meanings of MD in our data set. Other than SR, what could be the cause of shape variation? It probably has many causes, certainly among which is species evenness (Deline, 2009), but also functional and ecological attributes (Roy & Foote, 1997), environmental conditions (Vasil'ev et al., 2010), founder effects (David, 1999; Whitlock & Fowler, 1999), endemicity (Magniez-Jannin et al., 2000) or reproduction mode (Baltanas et al., 2002). The only factor we could discuss with our data was the habitat, an environmental parameter to which insect metric properties are known to be sensible (Benítez et al., 2011; Tantowijoyo & Hoffmann, 2011). Insects were collected in the forest, in rural and urban areas. It was difficult to address the question of habitat influence for a given species since the within species sample sizes did not allow valid statistics inference (see Table 1). Thus, the habitat as explored here was not free from the possible effects due to different species compositions. Our ANOVA analyzes could however provide some indications. When both species richness and habitat heterogeneity were significantly contributing to the MD, a good relationship between MD and SR was not observed. A broad morphological covering (TNLM), or some unpredictive elementary units of morphology (LNLM configurations like 1-4-6-9 or 1-3-5-6), were under the influence of both species and habitat heterogeneities (Table

7). On the contrary, the highly predictive LNLM configurations (like 2-3-8-13 or 2-6-7-13) were apparently under the influence of SR only (Table 7).

5. Perpectives

Almost two decades ago, Foote (1993) claimed that " *discordances between morphological and taxonomic diversity demand to be interpreted biologically, not explained away as artifact of taxonomic practice* ". Morphological disparity is obviously under the influence of factors other than the mere species number, and we showed here for instance the likely influence of the mosquito habitat. However, this does not preclude the possibility to use shape variation as an indicator of species richness. We considered this possibility through decomposing wing shape into elementary components and comparing their respective relationships with taxonomic richness.

We suggest that the use of elementary units of shape (LNLM) could allow one to focus on a single factor, like species richness, and that in this regard the use of many characters (TNLM) has the inconvenience of mixing various effects, making a clear interpretation difficult. In a recent study showing good parallelism between SR and metric diversity, a single character was used (Dolan et al., 2006).

If the objective of the morphometric analysis was to accurately reflect one factor, then the use of LNLM is recommended, although not any LNLM configuration. The remaining question is: which LNLM configuration to use?

The answer to such a question certainly implies to explore the relationship of the LNLM configurations with known factors other than possible species richness, like the habitats, the localities of collection, etc. Any combination of landmarks which would vary under the influence of such parameters would be less likely to reliably reflect species richness alone. However, a more definitive answer cannot be provided through the use of a model which did not reflect natural conditions closely enough. As explained above, an investigation is still needed to evaluate the interference of species evenness on the relationship between MD and SR (Deline, 2009), an issue not contemplated in our model.

6. Conclusion

At this stage, our data confirmed that metric properties of a given community contain hidden but accurate information about species richness. Our model suggests that this information is likely to be found through the examination of some elementary shape configurations rather than of a global multivariate projection of many morphological traits.

7. Acknowledgment

We thank R. Rattanarithikul for her kind help in taxonomic identification, as well as P. Kittayapong and C. Apiwhatnasorn (Mahidol University, Bangkok, Thailand) who encouraged this work. This study has been supported by the TRF/BIOTEC Special Program for Biodiversity Research and Training Grant BRT R_352052 and the Commission on Higher Education, RMU5080060 (Thailand), as well as by the IRD grants number HC3165-3R165-GABI-ENT2 and HC3165-3R165-NV00-THA1.

8. References

Baltanas, A., Alcorlo, P. & Danielopol, D. L. (2002). Morphological disparity in populations with and without sexual reproduction: a case study in *Eucypris virens* (Crustacea : Ostracoda), *Biological Journal of the Linnean Society 75(1): 9–19.*

Benítez, H. A., Briones, R. & Jerez, V. (2011). Intra and inter-population morphological variation of shape and size of the chilean magnificent beetle, *Ceroglossus chilensis* in the Baker River basin, Chilean Patagonia, *Journal of Insect Science, 11(94): 1–9.*

Ciampaglio, C. N., Kemp, M. & McShea, D. W. (2001). Detecting changes in morphospace occupation patterns in the fossil record: characterization and analysis of measures of disparity, *Paleobiology* 27(4): 695–715.

David, P. (1999). A quantitative model of the relationship between phenotypic variance and heterozygosity at marker loci under partial selfing, *Genetics, 153: 1463–1474.*

Deline, B. (2009). The effects of rarity and abundance distributions on measurements of local morphological disparity, *Paleobiology* 35(2): 175–189.

Dolan, J. R., Jacquet, S. & Torréton, J.-P. (2006). Comparing taxonomic and morphological biodiversity of Tintinnids (planktonic ciliates) of New Caledonia, *Limnology and Oceanography, 51 (2)* pp. 950–958.

Foote, M. (1992). Rarefaction Analysis Of Morphological And Taxonomic Diversity, *Paleobiology* 18(1): 1–16.

Foote, M. (1993). Discordance and Concordance Between Morphological and Taxonomic Diversity, *Paleobiology* 19(2): 185–204.

Gaston, K. J. & Spicer, J. I. (2004). Biodiversity: An introduction, *Oxford: Blackwell Publishing.*

Magniez-Jannin, F., David, B. & Dommergues, J. L. (2000). Analysing disparity by applying combined morphological and molecular approaches to French and Japanese carabid beetles, *Biological Journal of the Linnean Society 71(2): 343–358.*

Moyne, S. & Neige, P. (2007). The space-time relationship of taxonomic diversity and morphological disparity in the Middle Jurassic Ammonite radiation, *Palaeogeogr. Palaeoclimatol. Palaeoecol., 248: 82–95.*

Navarro, N. (2003). MDA: a MATLAB-based program for morphospace-disparity analysis, *Computers & Geosciences, 29: 655–664.*

Rattanarithikul, R., Harrison, B. A., Panthusiri, P. & Coleman, R. E. (2005). Illustrated keys to the mosquitoes of Thailand I. Background; geographic distribution; lists of genera, subgenera, and species; and a key to the genera., *Southeast Asian J Trop Med Public Health, 36(Suppl 1): 1–80.*

Rattanarithikul, R., Harrison, B. A., Panthusiri, P., Peyton, E. L. & Coleman, R. E. (2006). Illustrated keys to the mosquitoes of Thailand III. Genera *Aedeomyia, Ficalbia, Mimomyia, Hodgesia, Coquillettidia, Mansonia*, and *Uranotaenia.*, *Southeast Asian J Trop Med Public Health, 37(Suppl 1): 1–85.*

Rattanarithikul, R., Harrison, B., Harbach, R., Panthusiri, P. & Coleman, R. (2006.). Illustrated keys to the mosquitoes of Thailand IV. *Anopheles.*, *Southeast Asian J Trop Med Public Health 37(Suppl.2): 1–128.*

Rattanarithikul, R., Harbach, R. E., Harrison, B. A., Panthurisi, P., Coleman, R. E. & Richardson, J. (2010). Illustrated keys to the mosquitoes of Thailand. VI. Tribe Aedini, *Southeast Asian J Trop Med Public health, Vol. 41 (Suppl.1): pp. 225.*

Rohlf, F. J. (1990). Rotational fit (Procrustes) methods, *in* F. Rohlf & F. Bookstein (eds), *Proceedings of the Michigan Morphometrics Workshop. Special Publiation Number 2.*

The University of Michigan Museum of Zoology. Ann Arbor, MI, pp380, University of Michigan Museums, Ann Arbor, pp. 227–236.

Rohlf, F. J. (1996). Morphometric spaces, shape components and the effects of linear transformations, *in* L. F. Marcus, M. Corti, A. Loy, G. Naylor & D. Slice (eds), *Advances in Morphometrics. Proceedings of the 1993 NATO-ASI on Morphometrics*, New York: Plenum Publ. NATO ASI, ser. A, Life Sciences, pp. 117–129.

Roy, K., Balch, D. P. & Hellberg, M. E. (2001). Spatial patterns of morphological diversity across the Indo-Pacific: analyses using strombid gastropods, *Proc. R. Soc. Lond., 268: 2503–2508*.

Roy, K. & Foote, M. (1997). Morphological approaches to measuring biodiversity, *Trends In Ecology & Evolution* 12(7): 277–281.

Shannon, C. E. & Weaver, W. (1949). The mathematical theory of communication, *University of Illinois Press, Urbana*.

Simpson, E. H. (1949). Measurement of diversity, *Nature, 163: 688*.

Tantowijoyo, W. & Hoffmann, A. A. (2011). Variation in morphological characters of two invasive leafminers, *Liriomyza huidobrensis* and *L. sativae*, across a tropical elevation gradient, *Journal of Insect Science, 11(69): 1–16*.

Vasil'ev, A. G., Vasil'eva, I., Gorodilova, Y. V. & Chibiryak, M. V. (2010). Morphological disparity in populations with and without sexual reproduction: a case study in *Eucypris virens* (Crustacea : Ostracoda), *Russian Journal of Ecology, 41(2): 153–158*.

Whitlock, M. C. & Fowler, K. (1999). The distribution of phenotypic variance with inbreeding, *Evolution, 53(4): 1143–1156*.

Wills, M. A., Briggs, D. E. G. & Fortey, R. A. (1994). Disparity As An Evolutionary Index - A Comparison Of Cambrian And Recent Arthropods, *Paleobiology* 20(2): 93–130.

Zelditch, M. L., Swiderski, D. L., Sheets, H. D. & Fink, W. L. (2004). *Geometric morphometrics for biologists: a primer*, Elsevier, Academic Press. New-York.

3

Reproduction and Morphohlogy of the Travancore Tortoise (*Indotestudo travancorica*) Boulenger, 1907

Nikhil Whitaker,
Madras Crocodile Bank Trust, Mamallapuram,
Tamil Nadu,
India

1. Introduction

The most threatened chelonians occur in Asia, where virtually all species are heavily harvested for food and traditional medicinal trades (van Dijk *et al.* 2000). This includes the two endemic chelonians, *Indotestudo travancorica,* and the sympatric *Vijayachelys silvatica* in the Region of the Western Ghats, South India. *Indotestudo travincorica* is listed as Vulnerable under the IUCN Red List, and comes under Schedule IV of the Indian Wildlife (Protection) Act.

Phylogenetic relationships between the three species of *Indotestudo,* these being *forstenii, elongata,* and *travancorica,* have been in flux. Iverson *et al* (2001) examined these clades, and found that *I.travancorica,* the species in question here, was found to be more closely related to *I. elongata.* They also place *Indotestudo forstenii* from Sulawesi and Halmahera, as a distinct species unrelated to *I.travancorica* and *I. elongata,* refuting information that *I. forstenii* were introduced from India to Indonesia.

In 1982, 14 sub-adult *Indotestudo travancorica* were collected from Kerala (10° 5, 76°4), by the then MCBT (Madras Crocodile Bank Trust) researcher J. Vijaya, who also worked on *Vijayachelys silvatica* (Whitaker & Jaganathan, 2009). Between 1988 - 1995, the captive group of Travancore tortoises (*Indotestudo travancorica*) at the MCBT have produced 21 clutches of eggs (Whitaker and Andrews, 1997). However, not much information resulted from these clutches, as most of them were infertile (but see below). Apart from this publication, there are only a small number of other publications on the biology of this species, by Appukuttan (1991), Ramesh (2002, 2003, 2007, 2008), Bhupathy and Choudhury (1995), Vasudevan et al (2010), and Vijaya (1983). Here I examine factors related to the reproductive biology, morphology, and temperature selection in the species, in between the years 1999 - 2001, and 2008 - 2011.

2. Methods

The adult *Indotestudo travancorica* breeding pen at the MCBT houses two males, fifteen females , and two sub-adults. It consists of two interconnected exclosures, measuring 105 sq meters. One of the two exclosures was meshed off for a *Geochelone elegans* enclosure, on 18th

December 2010, which left a single exclosure for *Indotestudo travancorica*, reducing area to 87 sq meters. The exclosures contain *Pongamia* and *Bambusa*, shade trees. Leaf litter is removed and replaced every six months. A circular pond 100 centimeters in diameter varies from 5 - 10 cm in depth, and this provides drinking water and heat sinks. Hatchlings are housed in a separate outdoor terrarium measuring 2.54 x 0.55 x 0.28 m. The substratum is beach sand covered with dry bamboo leaf litter, and both substrata are changed on a regular basis. A bowl of water is provided filled with pebbles leaving 2 - 4 cm exposed, to prevent accidental drowning. V-shaped roof tiles are used as shelters.

Juvenile and adult tortoises are fed on tomato, carrots, beans, and various types of spinach, pumpkin, grasses, and beef. Food is placed at three feeding stations, on granite slabs measuring 1.2 x 1.2 meters, and 5 centimeters off of the exclosure floor. This occurs between 1000 - 1030 hrs, and 1530 - 1600 hrs, and any remnant feed is removed the following morning/evening Precise quantities of feed offered are not recorded. Feed is provided every day.

Chelonian embryos are at the gastrula stage (at a presomite stage of development) at oviposition (Ewert, 1979), as compared to crocodilians, with 16 - 18 somites at oviposition (Ferguson, 1985). The most advanced eggs of the Reptilia at egg laying are the lepidosaurs with 20-30 pairs of somites at oviposition (Muthukkaruppan *et. al.* 1970).Eggs are candled utilizing the technique used to candle crocodilian eggs as described by Hutton & Webb (1990), under a focused light, to determine the presence or absence of sub-embryonic fluid. This is found to be inaccurate in *I. travancorica,* and eggs deemed non viable hatch and vice versa (see section 3.3) Part of the chalking process in turtles, the development of an opaque band, involves adhesion of the vitelline membrane to the shell membrane, and this determines to some degree subsequent embryonic orientation (Ewert 1985). In addition, Andrews and Mathies (2000), note that adhesion of the embryo to the egg shell following oviposition may have a respiratory purpose "since chalking (drying) of the shell that is associated with adhesion increases the conductance to gases". However, in the travancore tortoise, viability of eggs is best confirmed by the presence of vascularisation under a candling lamp around a month into incubation, and by embryonic movement later in incubation.

Eggs were collected from 22 nests laid between December 1999 and September 2011 (Table 1). Eggs and hatchling tortoises were measured with vernier calipers (±0.1 mm) and weighed with an ACCULAB ™electronic weighing scale (±1 gram resolution). Eggs were then candled for presence or absence of sub-embryonic fluid. Clutches were segregated into different boxes, half immersed in vermiculite media, and allowed to develop at ambient room temperature. When egg laying was observed, female morphometrics, namely carapace length (mm), carapace width (mm), plastron length (mm), and weight (grams) were measured after females had compacted the nest. Where sexual size dimorphism in adults was done, four male and eighteen females were used from both the current MCBT population and animals measured in-situ by the late MCBT researcher J. Vijaya. Two females were included from a study by Bhupathy & Choudhury (1995), and nineteen females were from a dataset collected by Arun Kalagaven, from Eastern Kerala. The online database of the Global Diversity Information Facility revealed 26 specimens of unknown sex/size residing at the Florida Museum of Natural History (GBIF, 2011). Locations and GPS coordinates are not given on purpose, to protect in-situ populations.

Incubation temperature was monitored utilizing a HOBO XT ™ automatic temperature logger (Onset Instruments, P.O. Box 3450, Pocasset, MA 02559, USA) set to record temperature 4 times a day throughout the incubation period, at 800, 1200, 1600, and 2000 hrs for a group of 11 clutches laid in 1999. To get mean temperatures of all of the incubation boxes, the thermocouple from the logger was positioned in the middle of all boxes. Eggs were placed in 10 centimeter by 8 centimeter Tupperware boxes, within a plywood cupboard, to avoid predation of eggs by *Rattus bengalensis*. Incubation substrate was fine grain vermiculite. In addition, a bottle of beer with ca. ¼ of the contents left, and a piece of small beef within, was used in 2009 - 2011 to provide additional protection of eggs from fruit flies, as described by Wolff (2007) When oviposition was observed, eggs were removed from nests following compaction of the nest by the female.

Eggs that have rotted or cracked were discarded to avoid attracting flies and ants to other eggs in the same box. This was clearly evident from loss in egg weight and/or a sulphur-like smell. Morphometric measurements were recorded once hatching occurred, these being carapace length (millimeters), carapace width (millimeters), plastron length (millimeters), plastron width (millimeters), and weight (grams). A "successful hatching" was defined as an event where a hatchling managed to "pip", emerge from the egg, and survive >7 - 10 days post hatch This was the number of days typically required to absorb external yolk from pre-mature hatchlings. Hatchlings were measured and weighed within 12 hours of hatching, residual yolk outside the abdominal cavity varied from small, normal quantities, to abnormally large yolk sacs, with these hatchings not surviving.

To record tortoise temperature, an I-button TM was lodged on top of the carapace, and affixed with Gorilla glue ™. At between 30 - 90 days, animals were captured, and temperatures for the preceding days were recorded via the USB interface provided with the I-button software/I-buttons onto a Lenovo ™ laptop. Resolution of the I-buttons was at ± 0.5 °C. One male and four female tortoises was were measured for a period of 400+ days, and a sample is given here for 157 days, resulting in 943 observations (temperature loggers were programmed to record temperature every six hours), between 30th May - 3rd December 2011. The number of observations per individual varies, as loggers fell off of animals. Searches in the leave litter had to be done to retrieve lost loggers. Temperatures for these periods were eliminated.

Statistical analysis follows that of McGuinness (1999) and Fowler et al. (1998). In addition, SPSS 10.0 was used to confirm manual calculation of statistics. The level of significance for all analyses was $P < 0.05$.

3. Results & discussion

3.1 Clutch and egg sizes of *Indotestudo travancorica*

Clutch size of *Indotestudo travancorica* clutches collected ranged from one to six eggs. Clutch sizes and morphology of eggs is presented in Table 1. One particularly large egg from a clutch laid on the of 9th September 2011 measured 152.39 mm (egg length), 123.41 mm (egg width), and weighed 85 grams. Sane & Sane (1988) hatched a single *Indotestudo travancorica*, incubation period for this one egg was 139 days. Das (1995) puts the average incubation period of this species at 146-149 days. Incubation period for the seven hatched *Indotestudo travancorica* in 1999 ranged from 128 - 159 days (X = 141.5 days), with temperatures ranging between 22.4 °C - 28.7 °C. All clutches with more than one egg had hatchlings emerging on different days.

Clutch # / Date	X egg length (mm)	X egg width (mm)	X egg weight (gms)	Clutch size	Total clutch weight (gms)
01 : 7th December 1999	44.73	37.08	36.03	3	108.1
02 : 7th December 1999	47.4	36.7	38.1	2	76.2
03 : 7th December 1999	46.47	37.5	38.07	3	114.21
04 : 7th December 1999	48.47	38.77	43.43	3	130.29
05 : 2nd December 1999	45.77	35.98	34.93	3	104.79
06 : 4th December 1999	51.56	39.07	44.78	4	179.12
07 : 8th December 1999	42.23	36.18	33.05	2	66.1
08 : 8th December 1999	53.1	42.38	57.1	2	114.20
09 : 8th December 1999	45.3	36.53	36.13	3	108.39
10 : 14th December 1999	44.39	37.5	36.25	4	145.0
11 : 26th December 1999	50.48	39.62	47.33	3	141.99
12: 8th November 2010	51.5	40.7	51.7	3	155
13: 8th November 2010	48.4	37.2	37.0	2	74.0
14: 8th November 2010	49.9	38.6	44.0	1	44
15: 8th November 2010	52.4	39.2	46.8	5	234
16: 4th January 2011	48.7	37.1	42.7	6	256
17: 21st January 2011	50.3	38.5	44.9	2	90
18: 1st March 2011	48.2	33.2	44.1	2	88.10
19: 2nd March 2011	51.7	38.3	46.7	2	93.40
20: 22nd March 2011	50.8	37.4	44.7	3	134
21: 9th September 2011	84.0	67.3	59.7	3	179
22: 22nd September 2011	49.6	39.7	48	2	96
23: 1st October 2011	50.48	39.78	48	2	96
Average S.D. (Range)	50.26 ±7.89 (42.23- 84)	39.32 ±6.38 (33.2-67.3)	43.63 ±6.89 (33.05-59.7)	2.83 ±1.11 (1-6)	122.96 ±51.3 (44 – 256)

Table 1. Clutch sizes, morphometrics of eggs, of *Indotestudo travancorica* nests at the Madras Crocodile Bank.

Hatching success was 63.6 % from eggs laid in 1999, and it was not possible to incorporate eggs laid post this period as they are currently undergoing incubation. Interesting to note is that Viarda (2003), who maintains a captive colony of the sister species, *Indotestudo forstenii,* had clutch sizes ranging between one and two eggs. The same author noted that incubation period was between 101 – 130 days for this species, which overlaps with incubation periods of *Indotestudo travancorica.*

3.2 Female size, clutch correlates and notes on related behavior

I found six nests associated with females, as hole digging (Ramesh 2002) or oviposition was in progress. All nests were observed being dug post 1600 hours in all cases, save for one laid on 13th March 2006, laid at 1030 hrs. Results are presented in Table 2. It is noted in that in this study clutch size ranged from 1 – 6 eggs, the clutch with 6 eggs appears to be the largest clutch observed. Vasudevan *et. al.* (2010) summarized clutch size, and reported a range of 1 – 5 eggs.

In all cases, female weight post oviposition was not related to clutch size (Pearson Correlation= -2.86, Significance = 0.582, N=6), Figure 1. Overall, clutch masses represented 4 % of total female weight. No information was available on inter-nesting intervals, or which females were the most fecund. Clutch size was not related to female carapace length (ANOVA, F=1.95, d.f=4, P=0.257). Seasonality of oviposition varies highly in *I. travancorica*, with eggs being laid in January, March, September, October, and December. It appears that this species, with its tropical climate, may lay twice a year, once in the wet season, (October - February), and once in the dry (March - September).

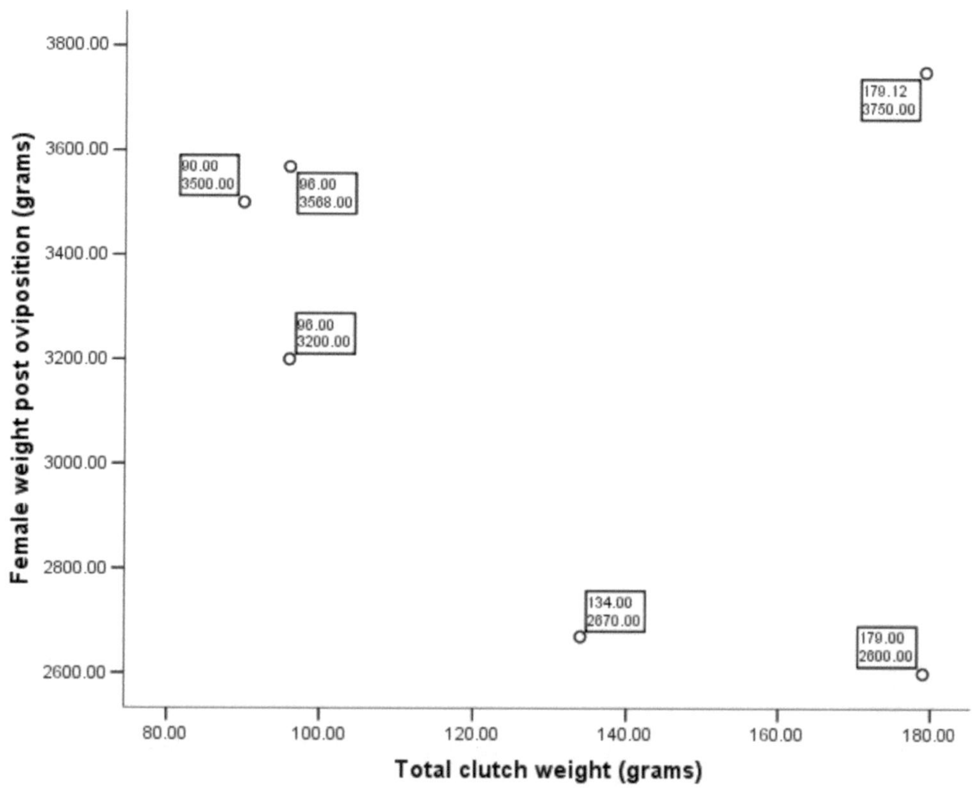

Fig. 1. Clutch weight (above text box, in grams) related to female weight post oviposition (grams).

Females were in a trance like state whilst digging/laying eggs (Plate 1), up until the time when they were packing the nest. Human observers and other tortoises were ignored during this period. In one instance, a female that laid on 20th September 2011, was noticed gathering/covering with *Bambusa* leaf litter at 1630, after which oviposition occurred, and the female was observed covering her nest post 1700. The nest was located at the periphery of a *Bambusa* clump. Following packing of the nest, she was observed to move directly towards the enclosure pond, and spent ca. one minute soaking herself and drinking water. She then went to one of the feeding stations and began to gorge on carrots, grass, and pumpkin with two other females.

Date laid	Female CL (mm)	Female CW (mm)	Female PL (mm)	Female Wt (gms)	Clutch size	X egg Length(mm)	X egg width (mm)	X egg weight (gms)	Total clutch wt (gms)
4th December 1999	290	250	200	3750	4	51.6	39.1	44.8	179.12
21st January 2011	261	161	208	3500	2	50.3	38.5	44.9	90
22nd March 2011	-	-	-	2670	3	50.8	37.4	44.7	134
9th September 2011	250	157	191	2600	3	84	67.3	59.7	179
22ndSeptember 2011	257	163	207	3200	2	49.6	39.7	48	96
1st October 2011	268	171	222	3568	2	50.48	39.78	48	96

Table 2. Relationship between clutch and female morphometrics in *Indotestudo travancorica.*

Plate 1. Oviposition in *Indotestudo travancorica*. Photograph by Shakti Sritharan, © Madras Crocodile Bank Trust.

Plate 2. *Indotestudo travancorica* female packing the nest; alternate hind limbs were used to do this. Photo by Nikhil Whitaker © Madras Crocodile Bank Trust.

Plate 3. Female *I. travancorica* were marked with a permanent marker once they start packing the nest; tortoises were easily identified and measured and weighed the following day. Photo by Nikhil Whitaker © Madras Crocodile Bank Trust.

3.3 Monitoring embryonic development

Of the four types of developmental arrest occurring in turtles described by Ewert (1985), embryonic diapause may be the situation in *Indotestudo travancorica* eggs. Once shifted from "natural" nests to the laboratory it was possible that humidity and temperature regimes might have changed to more favorable regimes for the species. Ewert (1985) defines embryonic diapause as "early developmental arrest at normal temperatures, irregular duration of this arrest, and the occasional detrimental effects of continuous normal development temperatures". Successful "natural" incubation within the *Indotestudo travancorica* enclosure is known. However, the time taken for eggs with unknown oviposition dates to hatch are similar to incubation periods described for the species. An alternative to the diapause theory was that eggs are laid shortly before collection. Another testudine, *Geochelone pardalwas*, is known to have diapause-like development (Carincross & Greig, 1977).

Chalked eggs in this study of *Indotestudo travancorica* appeared to indicate dehydration of the egg to the point where the embryos succumbed. Chalking was neither evident in eggs observed within the incubation boxes, or when viewed in front of a focused beam of light during the candling sessions. Given the importance of chalking, with regards to embryonic orientation and its association with gas exchange within the egg (Ewert, 1985; Andrews & Maties, 2000), this deserves further research. What was evident during the candling sessions was the extent of the extra-embryonic membranes in normally developing eggs. However, out of a total of 34 eggs candled in the 1999 incubation sessions, 18 had sub embryonic fluid visible, whilst 16 did not.

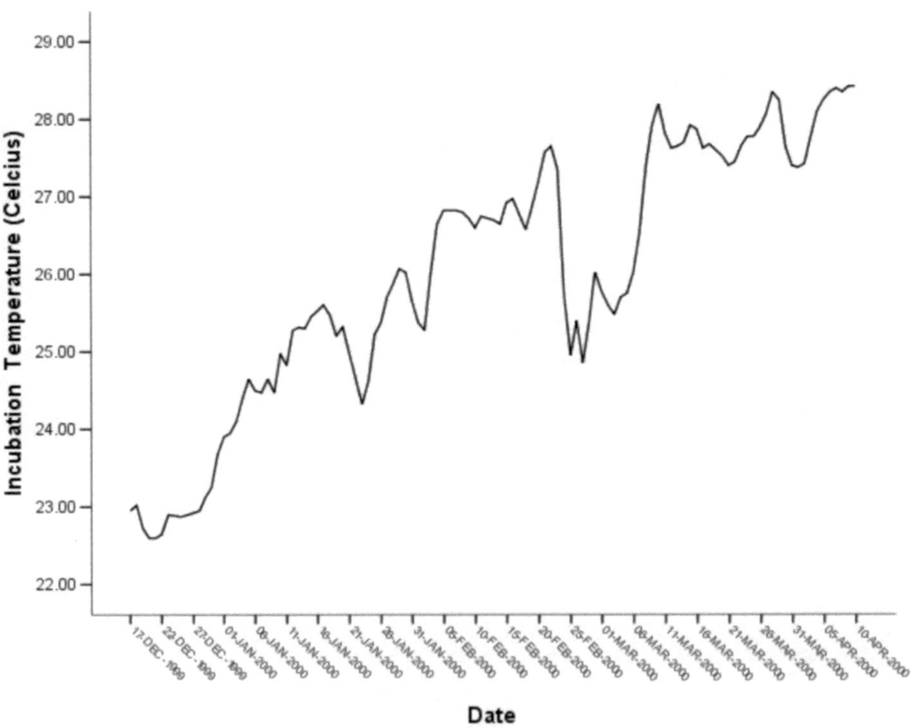

Fig. 2. Incubation temperatures for *Indotestudo travancorica* eggs incubated in 1999.

Air spaces were a common factor in *Indotestudo travancorica* eggs nearing termination of the incubation period, being confirmed by the afore mentioned candling technique, and they reflect a state of hydration (Ewert 1985). Ewert (1985) notes that air spaces were common in viable, brittle-shelled eggs, and recorded air spaces in *Rhinoclemmys annulata, R. wereolata,* and *R. punctularia,* evident from a few days before hatching. Metabolic heat produced by late stage embryos may promote water loss (Tracy, 1982). Figure 2 had six drops in average incubation temperature, and fourteen peaks; indeed in the last week of incubation, temperature became elevated to 28.4 ° C, from an average of 25.5 °C.

3.4 Diel variation in incubation temperature and incubation period

Temperatures at different times of the day, recorded for clutches between 1999 - 2000, are presented in Figure 3. At 800, temperature averaged 25.59, at 1200 25.92, at 1600 26.22, and at 2000 26.13 C; there is no significant variation in temperature between the 4 different times (ANOVA; F=3.19; p=0.032; d.f=467). Temperatures occasionally dropped drastically due to wetting of incubation media. Incubation temperature from the same group of eggs ranged from 22.4 ° C to 28.7 ° C (X=25.5° C; ±1.68 Figure 2). In a review of incubation times in reptiles, Birchard & Masseleni (1996) noted that mean incubation period for a group of 28 testudines averaged 30.0 ° C ±0.7, which was higher than the (25.5 ° C) mean I observed in this dataset.

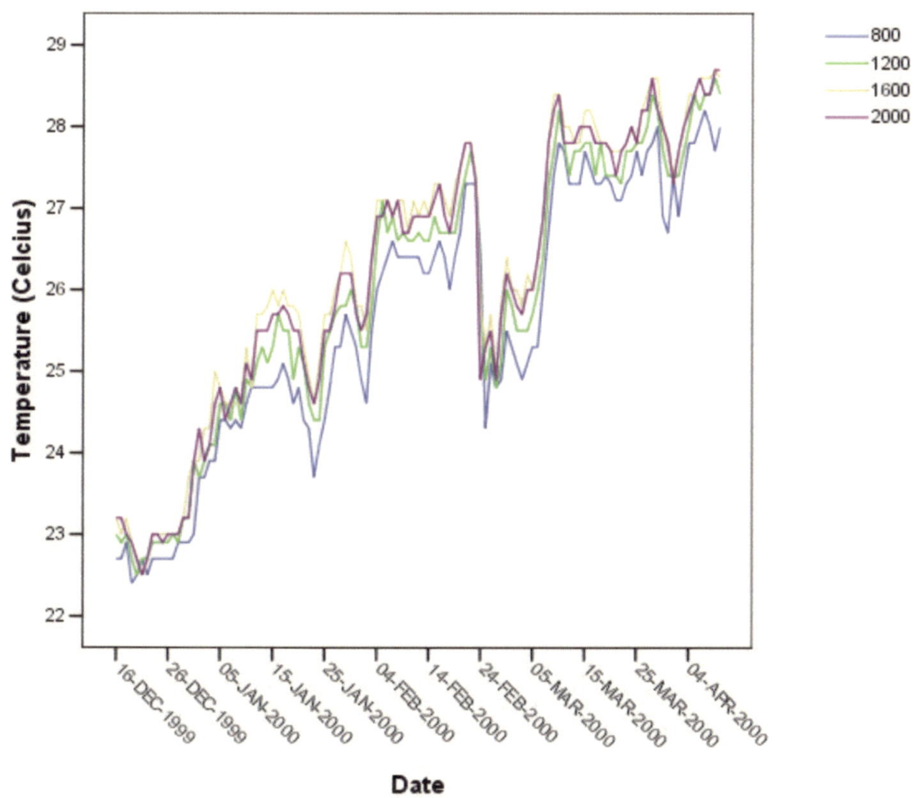

Fig. 3. Diel variation in incubation temperature of *Indotestudo travancorica* eggs incubated in 1999.

3.5 Hatchling morphometrics

One hatchling from eggs incubated in 1999 had a particularly large yolk sac and hatched 3 days premature to its single clutch mate, which hatched on day 128 of incubation and had an almost fully internalized yolk sac. The premature hatchling had the exposed yolk sac and umbilicus swabbed with Providine Betadine™, and was placed in a sterile container and then put into an incubator set at a constant 32.5° C in an attempt to speed up yolk internalization. However, the hatchling was found dead the next day.

Another hatchling with the same condition hatched 6th July 2011 (Plates 2 and 3) and had a much larger unabsorbed yolk sack. This animal died following 12 hours post hatch. Most individuals hatched with an external yolk sac, but these were rapidly absorbed within six to twelve hours (Plate4).

No.	CL (millimeters)	CW (millimeters)	PL (millimeters)	Hatchling weight (grams)	Initial egg weight (grams)
1	43.60	44.90	36.40	27.30	36.90
2	47.40	41.30	43.10	28.80	39.50
3	48.70	49.30	39.20	27.20	36.60
4	48.60	46.50	41.30	26.20	37.00
5	49.50	46.30	40.00	28.20	38.10
6	46.80	45.60	40.00	26.50	36.90
7	41.50	45.20	36.50	20.60	30.90
8	44.24	43.87	34.20	32.00	.
9	52.80	47.52	45.39	33.67	.
10	50.85	50.07	44.21	30.49	.
11	50.71	45.87	43.42	32.81	.
12	58.70	50.03	43.30	27.10	

Table 3. Hatchling morphometrics derived from clutches incubated in 1999 (with initial egg weight), and in 2011 (without initial egg weight).

Hatchling morphometrics, and relations to initial egg weight are presented in Table 3. Permeability of eggshells of *Chrysemys picta, Pseudemys concinna*, and *Trionyx muticus* almost doubles at mid to late incubation stages (Tracy 1982, in Ewert 1985). Average initial egg weight for seven hatchlings from 1999 was 10.2 gm heavier than hatchling weight. Average loss in percentage of initial egg weight as compared to hatchling weight was 72 %. This lies within Ewert's (1979) observations on juvenile turtles that were hatched from brittle shelled eggs, weighing 60 – 75 % of initial egg weight.

Plate 4. Premature *I. travancorica,* with a large external yolk. Photo by Nikhil Whitaker © Madras Crocodile Bank Trust.

Plate 5. Premature hatchling *I.travancorica.* Photo by Nikhil Whitaker © Madras Crocodile Bank Trust.

Plate 6. Plate 6. A normal, healthy *Indotestudo travancorica* with the yolk sac absorbed 24 hrs post hatching. Photo by Nikhil Whitaker © Madras Crocodile Bank Trust.

3.6 Sexual size dimorphism in *Indotestudo travancorica*

Sexes of *I. travancorica* did not differ significantly in any morphological attribute (Table 5). Specimens examined were from the Madras Crocodile Bank Trust's captive collection, and data collected within the archives of the MCBT by past researcher J.Vijaya at Nadukani, Kerala. Other data from the literature was gleaned from Bhupathy & Choudhury's (1995) work, and Arun Kanagavel, obtained from the Vazachal, Athirapally, and Chalakudy Forest ranges in Kerala during his surveys of *I. travancorica* in these areas.This agrees with Ramesh's (2008) work, wherein she examined 24 males and 25 females, and did not come to a consensus that sexes differed in carapace length. Interesting to note here, this is perhaps the first time individuals of *I.travancorica* from several locales have been combined for analyses of size.

Wilbur and Morin (1988), noted that if males were able to control copulation, such as in the MCBT's captive tortoise breeding enclosure, where copulation occurred regardless of interference by other males, then males will exceed females in body size and this allows them to forcibly inseminate females (plate 7). The variables introduced by maintaining *I. travancorica* in a captive environment as opposed to a natural population where they are free to choose diet, microhabitats, distance between animals, does not occur. Neither is there data available from the natural populations mentioned.

Plate 7. Mating in *Indotestudo travancorica*. Photo by Nikhil Whitaker © Madras Crocodile Bank Trust.

Sex	N	Parameter	X	SD	SEM	Range	Dimor-phism	F Crit	P
M	10	SCL	248.3	48.38	15.23	187-331	M=F	0.522	0.476
F	20	SCL	238.9	23.58	5.27	192-273	-		
M	10	CW	157.1	26.59	8.41	120-195	M=F	0.061	0.806
F	20	CW	159.26	19.91	4.57	135-225	-		
M	9	PL	182.89	34.91	11.64	139-224	M=F	0.359	0.554
F	20	PL	189.03	17.8	4.32	165-221	-		
M	9	MASS	2420.3	1470.37	490.12	876-4800	M=F	0.018	0.896
F	20	MASS	2036.8	729.55	167.37	1100-3600	-		

Table 4. Morphometric analysis was via ANOVA of adult *Indotestudo travancorica* from the Madras Crocodile Bank Trust's captive collection, data collected within the archives of the MCBT by past researcher J.Vijaya at Nadukani, Kerala , Choudhury & Bhupathy (1995), and Arun Kanagavel, collected from the Vazachal, Athirapally, and Chalakudy Forest ranges in Kerala.

2 male and 3 female *I.travancorica* were necropsied by J. Vijaya from the natural habitat of the species. CL for the 2 males averaged 154.5 millimeters (S.D. 6.36, 150 - 159), carapace width averaged 106.25 millimeters (S.D. 3.18, 104 - 108.5). The right testes length averaged 10.5 millimeters (S.D.3.54, 8-13), whilst the right testes width averaged 3.15 millimeters (S.D. 2.12, 2-5). The left testes length was 11.0 millimeters for both specimens.

With regards to the three females, CL averaged 190 millimeters (S.D. 22.07, 167 - 211), carapace width averaged 131 mm (S.D. 6.08, 124 - 135), plastron length for two females averaged 157.3 millimeters (S.D. 18.73, 144 - 170.5). In one of the necropsied females, (Carapace length, 192 millimeters, carapace width 135 millimeters, plastron length 170.5 millimeters, and weight 1100 grams) the right oviduct weighed 2.5 grams, whilst the left weighed 3 grams. Newly developed corpea lutea in the right oviduct weighed one gram. Older corpea lutea in the left oviduct weighed two grams, whilst in the right oviduct this was one gram. Oviducal eggs were 1 each in the left and right oviduct. This indicates that female *I. travancorica* reach maturity at between 192 - 290 millimeters, and around one kilogram in weight. (Table 4, J.Vijaya's information metioned above). A caveat here is that body size and age are not necessarily correlated (Justin & Gibbons, 1990).

The 1 male and 4 females from this study were measured on June 30th 2010, and the male measured 33 centimeters carapace length and weighed 5 kilograms. The four females averaged 24.34 centimeters carapace length, and 2900 grams. At the next measurement on 3rd August 2011, the male's carapace length was 32.4 centimeters, and weight was 4800 grams. The females averaged 26.73 centimeters carapace length , and 2350 grams in weight. The reduction in weight is real, whilst the reduction in carapace length by 0.6 centimeters was probably an observer error. Ernst & Barbour (1972), note, as observed in this study, that growth in turtles decreases following maturity, evident from the 1 year between measurements of captive animals here. Snider & Bowler (1992), record longevity of *Indotestudo travancorica* (then *I. forstenii*), as 26 years, however males and females in the MCBT captive group are approximately 30 years old.

The largest male from this study had a carapace length of 331 millimeters, and weighed 4800 grams, whilst the largest female had a carapace length of 271 millimeters and weighed 3600 grams. Both these tortoises were from the Madras Crocodile Bank's captive collection.

3.7 Temperature selection

The basic morphological versus temperature related parameters are presented in Table 5. The highest average temperatures were for the male, the heaviest animal considered here, 35.5 ° C. The smallest female's maximum average, female 3 was identical. No inferences can be made at this stage on this pattern. Both these animals also had the highest variance, inferring that they were perhaps the most active animals in this group. With regards to the male, he spent a large amount of the early morning and evening walking around the enclosure, in contrast to the females which had their movements largely restricted to the feeding stations.

Tortoise number and weight (grams)	Minimum	Maximum	Mean	Std. Deviation	Variance
Male 1 (4800)	19.5	35.5	28.01	2.31	5.33
Female 2 (2200)	20	33.5	27.36	2. 98	4.4
Female 3 (1900)	19	35.5	27.56	2.20	4.85
Female 4 (3300)	19.5	34	27.05	2.04	4.17
Female 5 (2000)	21.5	33.5	27.56	1.82	3.32

Table 5. Statistics relating to tortoise temperature (1 male and 4 females), between 30th May 2010 to 3rd December 2010.

Despite large size differences between the male and the females, temperature variation between the sexes was lower than expected (Figure 4). The male had the highest mean temperature 28.01 ° C, ranging between 21.5 - 35.5, and the highest variance (5.33 ° C). Consolidated female temperatures averaged 27.56 ° C , with an minimum and maximum of 20.0 ° C and 34.1 ° C respectively, and the highest maximum temperature at 34 ° C.

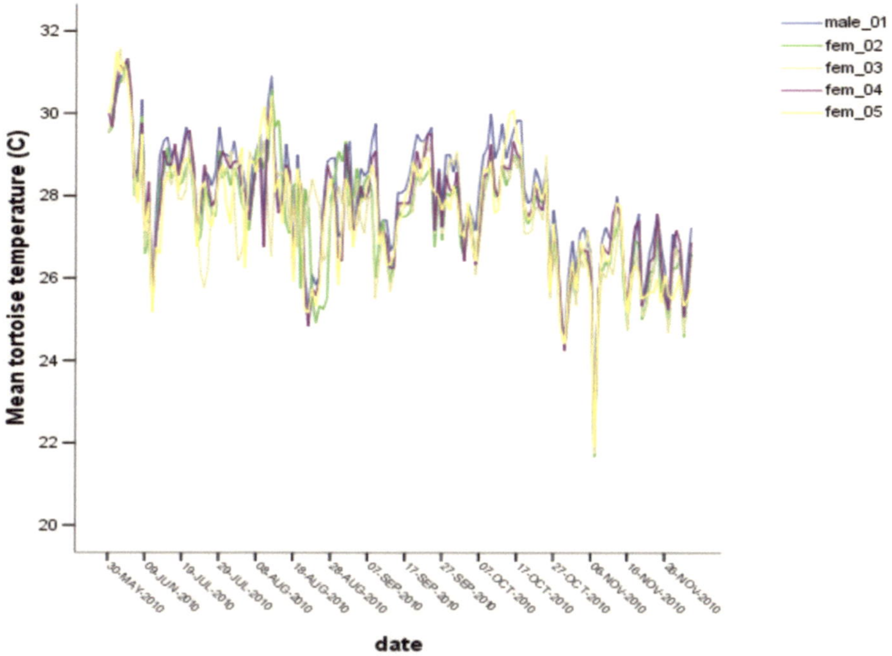

Fig. 4. Tortoise temperatures (1 male and 4 females) between 30th May and 3rd December 2010.

Females 2 and 5 had sudden drops in temperature, due to 5 millimeters of rainfall on November 5th 2010. Hailey & Coulson's (1995) observations on *Kinixys spekii* were similar to those observed in *Indotestudo travancorica*, in that during particularly hot days (i.e. > 35 ° C), movement was at a minimum, and on days that rainfall occurred, maximum activity was observed, this including courtship/mating, male combat, and nesting.

4. Conclusions

In this chapter I discussed the reproductive biology, morphometrics, and temperature selection in a small group of *Indotestudo travancorica*. Incubation period varied between 128 - 159 days (X = 141.5 days), in a group of 34 eggs from 11 clutches; out of these eggs candled in the 1999 incubation sessions, 18 had sub embryonic fluid visible, whilst 16 did not. From this series, diel variation did not differ between four measurements of temperature in a day. Initial determination of viability was not accurate, and definitive determination of viability did not occur until one month into incubation when vascularisation was visible.

Incubation periods were found to be similar in the sister species *Indotestudo forstenii*, but maximum clutch size was two, compared to six in *I.travancorica*. Clutch size was not related to female carapace length. Following egg laying , total clutch weight represented 4 % of the female's weight. Initial egg weight as compared to hatchling weight was 72 %. With regards to sexual size dimorphism, no differences were noted between males and females. *I. travancorica* females reach maturity at between 192 – 290 millimeters, and a kilogram in weight. Future studies may find females reaching maturity at smaller sizes. Temperature selection presented here in captivity as opposed to natural populations was a preliminary attempt at best; higher resolution loggers than the ones used here (±0.5° C) in combination with behavioral observations would result in a better data set.

5. Acknowledgements

Thanks to V. Sampath for his unfailing assistance in located *I.travancorica* clutches. I thank Romulus Whitaker for minding *Indotestudo travancorica* eggs while I was away and Madhuri Ramesh for valuable discussion. I thank Indraneil Das, Kartik Shanker, and Zai Whitaker for reviewing drafts of this manuscript. Thanks to P. Gowrishankar for providing bibliographical assistance. I wish to express my sincere thanks to Jeffery W Lang for donating the automatic temperature recorders used in 1999, and the Central Zoo Authority for their sponsoring of I-buttons used in temperature selection. I thank Arun Kanagavel, working under a Zoological Society of London, Erasmus Barlow Fellowship, for raw data on Travancore tortoise morphometrics. I am grateful to the Trustees of the Madras Crocodile Bank/Centre for Herpetology for allowing me to carry out this project, and access to archival data.

6. References

Andrews, R.M. & Mathies, T. (2000). Natural history of reptilian development: constraints on the evolution of viviparity. *Bioscience*. Volume 50, Issue number 3, pp. 227 - 238

Appukuttan, K.S. (1991). Cane turtle and travancore tortoise. An unpublished survey report, Kerala Forest Department.

Bhupathy, S. & Choudhury, B.C. (1995). Status, distribution, and conservation of the Travancore tortoise *Indotestudo forstenii* in Western Ghats. *Journal of the Bombay Natural History Society*. Volume 92, pp. 16 - 21

Carincross BL & Greig J.C. (1977). Notes on variable incubation period within a clutch of eggs of the leopard tortoise (*Geochelone pardalis*) (Chelonia: Cryptodira: Testudinidae).*African Zoology*. Volume 12, pp. 255 – 256.

Congdon, J.D. & Gibbons, J.W. (1990). The evolution of Turtle Life Histories. In: L*ife History and Ecology of the Slider Turtle*, Gans, C, pp. 45 – 54. Smithsonian Institution Press. Washington, D.C.

Das I. (1995). *Color guide to the turtles and tortoises of the Indian sub-continent.* R & A Publishing. United Kingdom.

Ewert, M.A. (1979). The embryo and its egg: development and natural history. In: *Turtles: Perspectives and Research,* Harless, M & Morloch, H., pp. 333-413, John Wiley & Sons,New York.

Ewert, M.A. (1985). Embryology of turtles. In: *Biology of the Reptilia, Vol. 14 A.*: Gans, C, pp. 75 – 267. John Wiley & Sons, New York

Ferguson, M.W.J. (1985). Reproductive Biology and Embryology of the Crocodilians, In: *Biology of the Reptilia, Vol. 14 A.* Gans, C, pp. 330 - 491. John Wiley & Sons, New York.

Fowler J, Cohen L, Jarvis P. (1998). *Practical statistics for field biology, (2nd edition).* John Wiley & Sons. England

GBIF (accessed through GBIF data portal, Herpetology specimens, http://data.gbif.org/datasets/resource/182, on 14th October 2011)

Hailey, A. & Coulson, I. M. (1995). Habitat association of the tortoises *Geochelone pardalis* and *Kinixys spekii* in the Sengwa Wildlife Research Area, Zimbabwe. *Herpetogical Journal.* Volume 5, pp. 305-309

Hutton J.M. & Webb G.J.W. (1990). An introduction to the farming of crocodilians, In: *A Directory of Crocodile Farming Operations,* pp. 1 - 139, compiled by R. Luxmoore, IUNN, Cambridge, United Kingdom.

Iverson, J.B ., Spinks, P.Q, Shaffer, H.B., McCord, W.P., & Das, I. (2001). Phylogenetic relationships among the Asian tortoises of the genus *Indotestudo* (Reptilia: Testudines: testudinidae). *Hamadryad.* Volume number, 26, Issue number 2, pp. 272 - 275

McDoual, J. & Castellano, C. (1996): Husbandry and captive breeding of the Travancore Tortoise (*Indotestudo travancorica*) at the Wildlife Conservation Society. In : *Advances In Herpetoculture,* 1996, pp. 51-56

McGuinness KA (1999). SBI 209 Design and Analysis of Biological Studies; Manual. School of Biological, Environmental, & Chemical Sciences, Northern Territory University, Australia, pp. 65

Moll, E.O. (1989). *Indotestudo forstenii,* Travancore tortoise. In: *The conservation biology of tortoises,* Swingland R. & Klemens, M.W. Occasional Paper, IUCN/SSC Number 5, pp. 119 – 120

Muthukkaruppan, V., Kanakambika, P., Manickavel, V. & Veeraraghavan K. (1970). Analysis of the development of the lizard, *Calotes versicolor. Journal of Morphrology.* Volume 130, pp. 479 - 489

Ramesh M. (2002). Observations on Travancore Tortoise (*Indotestudo forstenii*) in captivity. *Reptile Rap (Newsletter of the Reptile Network of South Asia)* Volume 4, pp. 4

Ramesh, M. (2003). Microhabitat description, morphometry and diet of the Travancore tortoise (*Indotestudo travancorica*) in the Indira Gandhi Wildlife Sanctuary, Southern Western Ghats. Masters Dissertation. Salim Ali School of ecology and environmental sciences, Pondicherry University, Pondichery, pp. 18.

Ramesh, M. (2007). Hole-nesting in captive *Indotestudo travancorica. Journal of the Bombay Natural History Society.* Volume 104, pp. 101.

Ramesh, M. (2008). Relative abundance and morphometrics of the Travancore tortoise, *Indotestudo travancorica,* in the Indira Gandhi Wildlife Sanctuary, southern Western Ghats, India. *Chelonian Conservation and Biology* Volume 7, pp. 108-113.

Sane L.S & Sane R.S. (1988). Some observations on the growth of the Travancore tortoise (*Geochelone travancoria*). *Journal of the Bombay Natural History Society.* Volume 86, Issue 1, pp. 109

Tracy, C.R. (1982). Biophysical modeling in reptilian physiology and ecology. In: *Biology of the Reptilia Vol. 12.* Gans C, Pough FH, pp. 275 – 321. New York: Academic Press.

Van Dijk, P.P., Stuart, B.L., Rhodin, A.G.J. (2000) Asian turtle trade: proceedings of a workshop on conservation and trade of freshwater turtles and tortoises in Asia. Executive summary. In: *Chelonian. Research. Monographs,* Volume 2, pp. 13-14.

Vasudevan, K., Pandav, B & Deepak, V. (2010) .Ecology of two endemic turtles in the Western Ghats. *Final Technical Report,* Wildlife Institute of India 74p.

Vijaya, J. 1983. The Travancore tortoise, *Geochelone travancorica. Hamadryad,* Volume 8, pp. 11-13.

Whitaker, N & Jaganathan. V. (2009). Biology of the forest cane turtle, *Vijayachelys silvatica,*in South India. *Chelonian Conservation Biology,* Volume 8, pp. 109-115.

Whitaker, R. & Andrews, H.V. (1997). Captive breeding of Indian turtles and tortoises at the Centre for Herpetology/Madras Crocodile Bank. In: *Proceedings of Conservation, Restoration, and Management of Tortoises and Turtles.* New York Turtle and Tortoise Society, New York. pp 166 – 170

Wilbur, H.M. & Morin, P.J. (1988). Life history evolution in turtles In: *Biology of the Reptilia. Vol. 16b.* Gans,C. & Huey, R,. pp. 396-447. Alan R. Liss, New York.

Wolff, B. 2007. The hump backed fly (*Phoridae*) as a risk to successful incubation. *Radiata* Volume 16, Issue 3, pp. 49 - 53

4

Morphometrics and Allometry in Fishes

Paraskevi K. Karachle and Konstantinos I. Stergiou
Aristotle University of Thessaloniki, School of Biology, Department of Zoology, Laboratory of Ichthyology
Greece

1. Introduction

Fish morphometrics has been in the hot-spot of ichthyological studies for many decades, but the initial steps date back to the time of Galileo Galilei (Froese 2006). Yet, the scientific basis for morphometry in fishes, and especially the mathematical way that weight relates to length, was set by Fulton, in 1906, who for the first time introduced fisheries science into 'allometry' (Froese 2006).

Nowadays, the most commonly used relationships, that have been established for the majority of fishes (Binohlan & Pauly 2000, FishBase: www.fishbase.org: Froese & Pauly 2011), are those relating weight to body length (in the majority of cases, total body length (TL)), and different types of length (i.e., standard (SL) and fork (FL) length) to TL. Weight (W)-length (TL) relationships are of power type, i.e., $W=a\,TL^b$. In this equation, a is the coefficient of body shape (Lleonart et al. 2000, Froese 2006, www.fishbase.org), and it gets values around 0.1 for fishes which are small sized and with a rounded body shape, 0.01 for streamlined-shaped fishes and 0.001 for eel-like shaped fishes. In contrast, b is the coefficient balancing the dimensions of the equation and its values can be smaller, larger or equal to 3 (Lleonart et al. 2000, Froese 2006, www.fishbase.org). In the first two cases (i.e., $b<3$ and $b>3$) fish growth is allometric (i.e., when $b<3$ the fish grows faster in length than in weight, and when $b>3$ the fish grows faster in weigth than in length), whereas when $b=3$ growth is isometric. Froese (2006) analyze 3929 weight-length relationships for 1773 species, and reports that b ranges between 1.96 and 3.94, with 90% of the cases falling inside the 2.7-3.4 range. The lowest values have been recorded for *Cepola macropthalma*, whereas the highest for *Chaenocephalus aceratus*. In principle, these types of relationships are allometric (82%), with a trend towards positive allometry (Froese 2006). Weight-length relationships are of high importance for fisheries science and can be used in a wide range of applications, such as: (a) estimation of biomass from length data; (b) estimation of a species condition factor; and (c) comparisons among life history and morphologic differentiations of the same species in different areas (e.g., Pauly 1993, Petrakis & Stergiou 1995, Binohlan & Pauly 2000).

In recent years, attempts have been made to relate other morphological characteristics of fishes, such as mouth (e.g., Karpouzi & Stergiou 2003, Chalkia & Bobori 2006, Karachle & Stergiou 2011a), intestine (e.g., Kramer & Bryant 1995a, b, Karachle & Stergiou 2010) and tail (Karachle & Stergiou 2004), to TL, and as well as to feeding habits and fractional trophic levels (τ). In general, eco-morphological studies focus on the patterns that relate morphology and the use of available resources (e.g., Motta et al. 1995, Wainwright & Richard 1995), and consider morphology as a key factor for the determination of a species' feeding habits. Hence, variations in morphology are due to differences in the ability of different fish species to catch and consume their food, affecting the overall diet composition (e.g., Wainwright & Richard 1995, Wootton 1998).

Mouth gape has long being considered as the most important, yet restraining, factor affecting food consumption mainly in: (a) defining the size range of prey items a consumer can catch/consume and (b) affecting the efficiency of a predator to catch and consume its food (Wainwright & Richard 1995). More specifically, mouth gape can be used for the evaluation of the relationship between prey and predator size (e.g., Keast & Webb 1966, Wainwright & Richard 1995), whereas mouth shape and position, teeth, structure and number of gill rakers seem to be related to the type of food being consumed (e.g., Al-Hussaini 1947, Kapoor et al. 1975, Verigina 1991). The size spectrum of prey items for fishes increases as they grow, which is more evident in apex predators (Karpouzi & Stergiou 2003), and this fact has been mainly attributed to ontogenetic changes related to mouth morphology, visual acuity, more efficient digestion and better swimming ability of large fish (e.g., Keast & Webb 1966, Kaiser & Hughes 1993, Juanes 1994, Juanes & Conover 1994, Hart 1997, Wootton 1998, Fordham & Trippel 1999). Hence, mouth morphometry is generally related to τ (Karpouzi & Stergiou 2003).

Intestine (or gut) length (GL) is considered to be an indicator of diet (Kramer & Bryant 1995a) and, particularly in fishes, can be used for interspecific dietary comparisons (e.g., Al-Hussaini 1947, Karachle & Stergiou 2010a). For a given body length, intestine in herbivorous species is longer than in omnivorous ones, and in omnivorous species longer than in carnivorous ones (e.g., Kapoor et al. 1975, Kramer & Bryant 1995b, Karachle & Stergiou 2010a, b). Hence, the widely accepted pattern of fish GL variation in relation to species feeding habits is:

Carnivores<omnivores<herbivores<detritus feeders

(e.g., Kapoor et al. 1975, Ribble & Smith 1983, Kramer & Bryant 1995b, Karachle & Stergiou 2010a, b). The same pattern is also true in other vertebrate classes (e.g., reptiles: O'Grady et al. 2005, birds: Ricklefs 1996, and mammals: Chivers & Hladik 1980).

Recent research has shown that there is a strong relationship between GL and body length (BL), that can be best described by the power type equation, i.e., $GL=a\ BL^b$ (Kramer & Bryant 1995a, b, Karachle & Stergiou 2010a, b). The significance of this allometry in GL could be related to the effect of the increasing body length to the relative efficiency of intestine to absorb nutrients from the digested food (e.g., Ribble & Smith 1983, Kramer & Bryant 1995a). Since growing organisms require more energy and nutrients, changes of the structural capacity, i.e. lengthier intestine, must be performed in order for those needs to be met. Those structural changes of intestine will ensure that food will be retained longer in the tract and more nutrients will be adsorbed, and more receptors for the absorption of energy and nutrients will be available.

Tail, the one characteristic that in the eyes of everyday people is what defines a fish, is the least examined of all feeding-related morphologic characteristics in fishes. Its relationship with feeding was established in a model of the estimation of annual food consumption per unit biomass, i.e., Q/B (Palomares & Pauly 1989). In this model, the tail aspect ratio (A; i.e., the ratio of squared tail height per tail area) is a key variable:

$$Q/_B = \frac{T^{0.61} \times A^{0.52}}{1.2 \times W^{0.2}}$$

where T is the mean water temperature (°C) and W is the asymptotic (or maximum) live weight of the fish (g) in the population (Pauly 1989a).

Despite the importance of tail characteristics, such as height, area and aspect ratio, little effort has been put into their study and thus available information is limited (Karachle & Stergiou 2004, 2005, 2008a). Moreover, even in research focusing on Q/B estimation, A has been estimated mainly from fish photos or drawings (e.g., García & Duarte 2002), or from measurements derived from a small number of individuals (e.g., 3-4 individuals; Angelini & Agostinho 2005).

In this chapter, we explore the effects of feeding habits, environment and/or habitat type on weight-length relationships and the morphometrics of feeding-related characteristics of fishes, namely mouth, intestine and tail. For weight-length relationships, mouth and intestine, we expanded on previously published information and relationships. For tail allometry we estimated tail area (TA), height (TH) and tail aspect ratio (A) for 61 species from the North Aegean Sea, Greece, using imprints of tails, for a large number of individuals per species. Based on these estimates, we established the within-species relationships between TA, TH and A with TL. We also explored the relationships of the above mentioned tail characteristics with species' feeding habits and habitat type (i.e., pelagic, benthopelagic, demersal, and reef-associated; the ecological niche concept being also included). Finally, the relationship between the mean A values (A_m) and mean TL (TL_m) per species was explored.

2. Materials and methods

We used prior published information on morphology and morphometrics of fishes, and especially mathematical expressions, which were transformed into allometric regressions and grouped based on species' feeding habits and environment/habitat. The use of the allometric model, instead of other types of models (i.e., linear, exponential, and logarithmic), in the description of relationships between morphological characteristics, such as those presented here, and body length, was better for the following reasons:

1. it allows the detection of morphologic changes in shape, which is the basic hypothesis in the study of morphometrics (Lleonart et al. 2002)
2. in contrast with the other types of models, the allometric one is the only one for which when X=0, then Y=0, a fact that is meaningful in morphometrics (Lleonart et al. 2002), and
3. especially in the case of feeding related characteristics, it is the only one that can explain changes in the morphology such as required for a growing fish to meet its increasing energetic demands, while the energy spent for the acquisition of food is minimized (Karachle & Stergiou 2010b, 2011a).

2.1 Weight-length relationships

In order to examine the effect of feeding habits and habitat type on the weight-length relationships, we used the data of Karachle & Stergiou (2008b) for 60 species from the North Aegean Sea. The individual data of all 60 species were plotted together in order to examine whether there is any pattern in the way that W changes with TL across species. Consequently, the weight-length data of the different species were grouped according to:

1. species' feeding habits, using their τ as given in FishBase (www.fishbase.org). Based on τ values, fishes were grouped into five different functional trophic groups (FTGs), using the classification of Stergiou & Karpouzi (2002):
   a. pure herbivores ($2.0 < \tau < 2.1$) (H);
   b. omnivores with preference for plants ($2.1 < \tau < 2.9$) (OV);
   c. omnivores with preference for animals ($2.9 < \tau < 3.7$) (OA);
   d. carnivores with preference for decapods/fish ($3.7 < \tau < 4.0$) (CD); and
   e. carnivores with preference for fish/cephalopods ($4.0 < \tau < 4.5$) (CC).
2. habitat type (i.e., pelagic, benthopelagic, demersal, and reef-associated; information from FishBase (www.fishbase.org)).

Next, the combined regression lines of species per FTG and habitat type were plotted on the same graph and patterns were identified. Finally, the combined general regression lines of W-TL relationships of each of the above mentioned groupings were compared, using the log-transformed data, with Analysis of Covariance (ANCOVA; Zar 1999).

2.2 Mouth characteristics

The relationships of mouth area (MA) with total body length (TL), which are given in Karpouzi & Stergiou (2003) and Karachle & Stergiou (2011a), were used. They were again grouped according to FTGs and habitat type (see section 2.1). Based on the original MA-TL equations for 68 species, general regression lines were constructed.

2.3 Intestine morphometrics

The relationships of fish gut length (GL) with body length (BL), presented in Karachle & Stergiou (2010a) and Karachle & Stergiou (2010b), were used in order to check for GL changes in relation to feeding habits and habitat. Overall, relationships of GL-BL were used for 99 species, and the individual data for these species were grouped according to the FTGs of the species and habitat (see section 2.1).

2.4 Tail

Samples were collected in the North Aegean Sea, on a seasonal basis, from June 2001 to January 2006, using commercial fishing vessels (i.e., trawlers, purse-seiners, and gill-netters) and preserved in 10% formalin (for details see Karachle & Stergiou 2008c). In the laboratory, TL was measured and tail was imprinted (for at least 30 individuals per species, when possible). Based on these imprints, tail height (TH) and area (TA) were estimated, using UTHSCSA IMAGETOOL Ver. 3.0 (Wilcox et al. 1997) software. Based on TH and TA measurements, A was estimated, as follows (Pauly 1989a):

$$A = \frac{TH^2}{TA}.$$

The relationships between TA-TL and TH-TL were established using power regression ($Y=a X^b$; Lleonart et al. 2000) and consequently b (given the mathematical traits of b as explained in the introduction) was tested for difference from 2, in the case of TA (since the measurement unit of TA is cm^2), and difference from 1, in the case of TH (since the measurement unit of TH is cm), using t-test (Zar 1999). Additionally, all TA-TL regressions were plotted together for the detection of possible groupings of species.

In order to identify patterns of changes of TA with TL, the data for the different species were compiled together based on FTGs and habitat type (see section 2.1). In order to identify possible patterns, the regressions per group in each of the above mentioned compilations were plotted together. Comparisons of the slopes of the general regression lines were performed on the log-transformed data using ANCOVA (Zar 1999).

The relationships between A and TL were estimated for all species (they are not presented here) and the type of the relationship was defined based on the R^2 values. Finally, the across-species relationship between the mean A values (A_m) and mean TL (TL_m) per species was also explored.

3. Results

3.1 Weight-length relationships

When all data for the 60 species were plotted together, three major groups were identified (Fig. 1a, b): group (I) included *C. macrophthalma* and *Belone belone*, group (II) *Scyliorhinus canicula* and *Sphyraena sphyraena*, and group (III) the remaining species. The slopes of the regressions of the three groups differed significantly (for all combinations: $p<0.05$). The graphs of the combined regression lines per FTG and habitat type (number of species per FTG and habitat type are given in Table 1) did not reveal any clear grouping of weight-length relationships (Fig. 1c and d). Nevertheless, in the case of FTGs, based on the results of ANCOVA, there was a significant difference in the slopes of the regression lines between omnivores with preference to animal material, carnivores with preference to fish and decapods and carnivores with preference to fish and cephalopods (for all combinations: $p<0.05$), whereas there was no difference between those of the regressions for herbivorous species with those of the species of all other FTGs (for all combinations: $p>0.10$). Accordingly, in the case of habitat type, there was not any significant difference in the slopes of the relationships between pelagic and benthopelagic species (ANCOVA: $p=0.569$, F-ratio=9026.64), while the slopes of all remaining combinations differed significantly (for all combinations: $p<0.05$).

For the same length, omnivores with preference to animal material weighed less than carnivores (carnivores with preference to fish and decapods and carnivores with preference to fish and cephalopods) (Fig. 1c). When habitat type was examined (Fig 1d), for the same length, the following pattern was observed for weight:

$$D<P<BP = RA.$$

Category	W	MA	GL
Functional Trophic Groups			
Herbivores (H)	1		5
Omnivores with preference to plants (OV)		1	5
Omnivores with preference to animals (OA)	33	36	55
Carnivores with preference to decapods and fish (CD)	8	9	16
Carnivores with preference to fish and cephalopods (CC)	18	22	18
Habitat type			
Pelagic (P)	15	15	17
Benthopelagic (BP)	14	14	29
Demersal (D)	25	29	45
Reef-associated (RA)	6	7	8
Total number of species	60	68	99

Table 1. Number of species per functional trophic group and habitat, for which weight (W), mouth area (MA) and intestine length (GL) relationships with body length were retrieved from the literature (W: Karachle & Stergiou (2008b); MA: Karpouzi & Stergiou (2003) and Karachle & Stergiou (2011a); GL: Karachle & Stergiou (2010a, b)).

3.2 Mouth characteristics

The distribution of the 68 species used for mouth morphometrics by FTGs and habitat is given in Table 1. The vast majority of species were omnivores with preference to animals, followed by carnivores with preference to fish and cephalopods. In the present study, among the 68 studied species, demersal species outnumbered those living in the remaining habitats, followed by pelagic and benthopelagic species, which were equally represented.

Plots of the regression lines of species per FTG revealed that for the same TL (Fig. 2a), MA dimensions change as followed:

$$OA < OV < CC < CD.$$

Comparison between the different habitats (Fig. 2b) did not reveal a clear pattern. Only benthopelagic species seemed to largely differentiate from the remaining three habitat-related groups, at lengths >20cm.

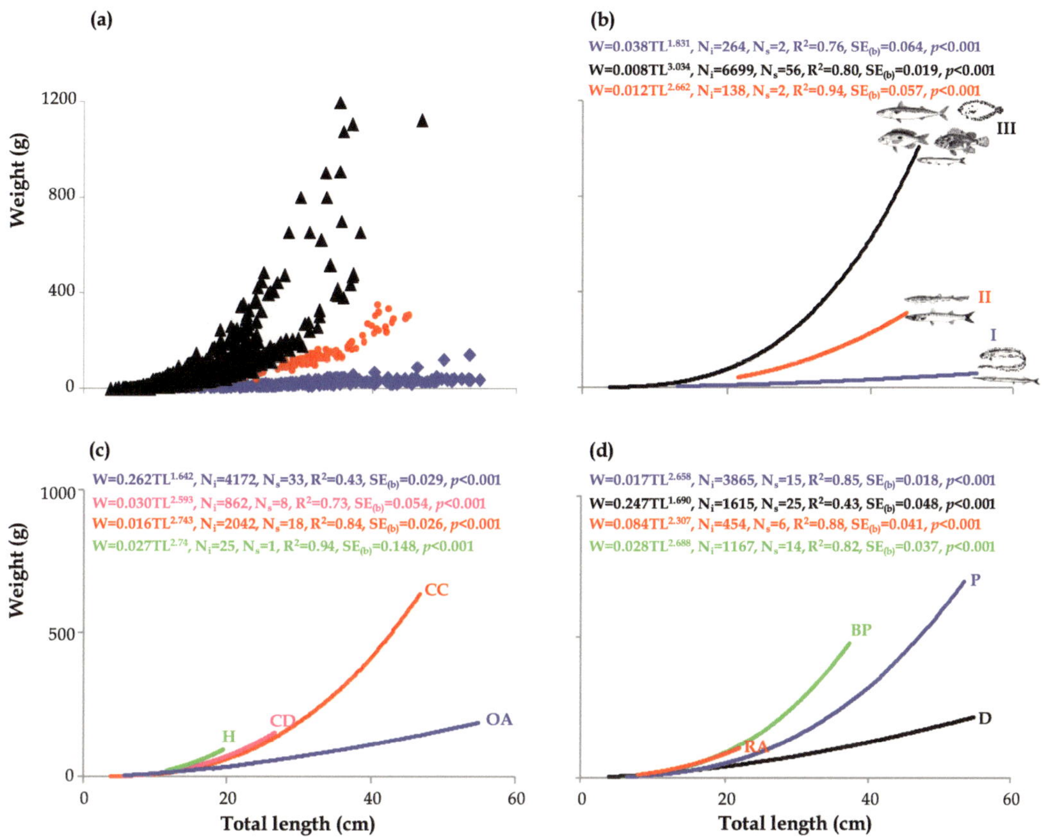

Fig. 1. Regressions between total body length (TL) and weight (W) based on data from Karachle & Stergiou (2008b) for 60 fish species North Aegean Sea, Greece, June 2001-January 2006: (a) original individual data, (b) groups identified in (a), (c) functional trophic group, as identified by Stergiou & Karpouzi (2002) and (d) habitat type (from Fishbase; www.fishbase.org: Froese & Pauly 2011). H: herbivores; OA: omnivores with preference to animal material; CD: carnivores with preference to decapods and fish; CC: carnivores with preference to fish and cephalopods; P: pelagic; BP: benthopelagic; D: demersal; RA: reef-associated; N_i=number of individuals; N_s=number of species; R^2 = coefficient of determination; and $SE_{(b)}$ = standard error of slope b. Fish drawings are from FishBase (www.fishbase.org; Froese & Pauly 2011).

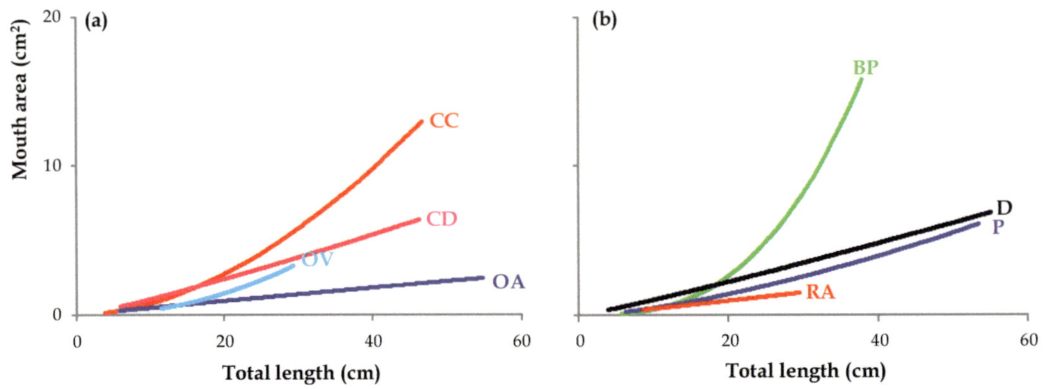

Fig. 2. Regressions between total length and mouth area based on literature data for 68 fish species (from Karpouzi & Stergiou (2003) and Karachle & Stergiou (2011a)) grouped by: (a) functional trophic group, as identified by Stergiou & Karpouzi (2002), and (b) habitat type. OV: omnivores with preference to vegetable material; OA: omnivores with preference to animal material; CD: carnivores with preference to decapods and fish; CC: carnivores with preference to fish and cephalopods; P: pelagic; BP: benthopelagic; D: demersal; RA: reef-associated.

3.3 Intestine morphometrics

Of the 99 different species for which GL-BL relationships were used, more than half (55 out of 99 species) were omnivores with preference to animals, and the vast majority of them were demersal species (Table 1).

The regression lines of the species per FTG showed a clear formation of two separate groups: one including herbivorous species and omnivores with preference to plants (group I), and another one including omnivores with preference to animal material, carnivores with preference to fish and decapods, and carnivores with preference to fish and cephalopods (group II) (Fig. 3a). Additionally, for the same BL, GL was higher for the species of the first group than in those of the second one.

Finally, GL changed with habitat type as follows (Fig. 3b):

$$D<BP \leq P<RA.$$

3.4 Tail

Overall, TA, TH and A values were estimated for 61 fish species (2703 individuals; Table 2). The number of individuals examined ranged from 6 (for *Lophius piscatorius* and *Pomatomus saltatrix*) to 100 (for *Arnoglossus laterna*) (Table 2). A_m ranged from 0.23, for *Gaidropsarus mediterraneus* and *Lesueurigobius suerii*, to 4.38, for *Scomber scombrus*.

The TA-TL and TH-TL relationships are shown in Table 2. They were all significant ($p<0.05$). In the case of TA-TL relationships, in 32 out of the 61 species (52.5%), the b value of the relationship was statistically different from 2 (t-test: $p<0.10$), indicating the predominance of the allometric relationship of TA and TL. This relationship was positively allometric (i.e., b>2)

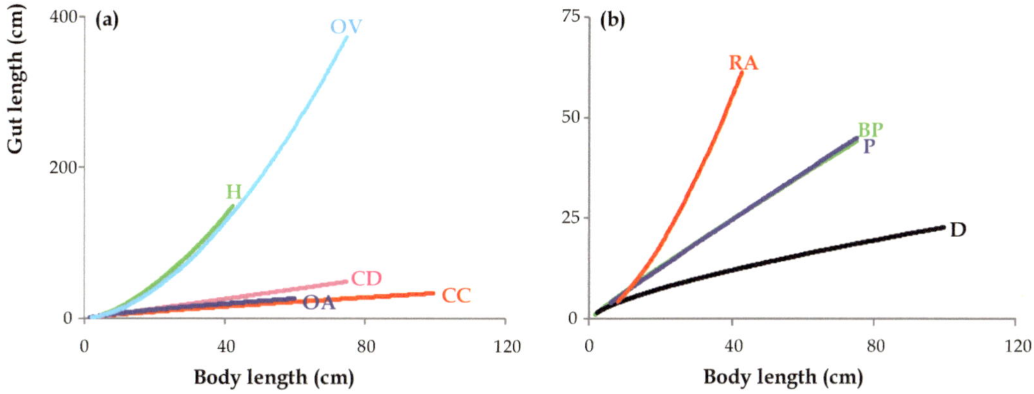

Fig. 3. Regressions between body length and gut length based on literature data for 99 fish species (from Karachle & Stergiou (2010a, b) grouped by: (a) functional trophic group, as identified by Stergiou & Karpouzi (2002), and (b) habitat type (from Fishbase; www.fishbase.org: Froese & Pauly 2011). H: herbivores; OV: omnivores with preference to vegetable material; OA: omnivores with preference to animal material; CD: carnivores with preference to decapods and fish; CC: carnivores with preference to fish and cephalopods; P: pelagic; BP: benthopelagic; D: demersal; RA: reef-associated.

in 20 species (62.5%), and negatively allometric (i.e., $b<2$) in 12 species (37.5%) (Table 2). For the remaining 29 species (47.5%) for which the power expression of the relationship was not statistically significant (t-test: $p> 0.10$), and based on the R^2 values, the relationship was of the power type for 14 species (48.3%), exponential ($TA = a\, e^{bTL}$) for 8 species (27.6%), linear ($TA=a+bTL$) for 5 species (17.2%) and logarithmic ($TA=a+b\log TL$) only for *Diplodus annularis* and *Diplodus vulgaris* (6.9%) (Tables 2 and 3).

Likewise, in the case of the TH-TL relationships, the allometric model was statistically accepted (i.e., $b\neq1$; t-test: $p<0.10$) for the majority of species (40 out of the 61 species; 65.6%). Positive allometry (i.e., $b>1$) was found for 35 species (87.5%), and negative allometry (i.e., $b<1$) for only 5 species (12.5%) (Table 2). For the remaining 21 species (34.4%) for which the assumption that $b\neq1$ was not statistically significant (t-test: $p>0.10$), the relationship type identified was: linear ($TH=a+bTL$) for 8 species (38.1%), power for 6 species (28.6%), exponential ($TH = a\, e^{bTL}$) for 5 species (23.8%) and logarithmic ($TH=a+b\log TL$) only for *Coris julis* and *Dentex dentex* (69.5%) (Tables 2 and 3).

When individual data for each species were grouped per FTG (Fig. 4), there was a clear separation of omnivores with preference to animal material, from carnivores with preference to fish and decapods, and carnivores with preference to fish and cephalopods. Indeed, the slope of the regression of omnivores with preference to animal material differed significantly from that of the other two regressions (ANCOVA: for both cases $p<0.01$), whereas there was no difference between the slopes of the regressions for carnivores with preference to fish and decapods and carnivores with preference to fish and cephalopods (ANCOVA: $p>0.10$). Accordingly, when habitat type was used for the grouping of species (Fig. 5), two separate groups were formed, one including reef-associated and benthopelagic species and another one including pelagic and demersal species and there was a significant difference among the slopes of all four regressions (ANCOVA: for all cases $p<0.01$).

Species	N	H	TL range (cm)		TL$_m$	TH - TL						TA-TL						Aspect ratio		
			min	max	(cm)	a	B	SE$_{(b)}$	R^2	b≠1	ToR	a	b	SE$_{(b)}$	R^2	b≠2	ToR	A$_m$	R^2	ToR
Alosa fallax	25	P	15	46.8	24.4	0.3634	0.9239	0.065	0.90		Li	0.0314	1.8886	0.083	0.96		P	3.71	0.026	Pn+
Anthias anthias	9	RA	12.7	16.6	14.6	0.0853	1.4855	0.505	0.55		P	0.0436	1.8632	0.279	0.87		L	3.29	0.204	Pn-
Apogon imberbis	30	RA	8	11.5	10.2	0.1842	1.2405	0.243	0.48		Li	0.0289	2.2014	0.281	0.69		E	2.25	0.053	Pn+
Arnoglossus laterna	100	D	4.5	16.6	11.4	0.1016	1.2101	0.030	0.94	×	P	0.0159	2.1737	0.040	0.97		P	1.18	0.397	Pn+
Belone belone	67	P	27.2	53.5	33.7	0.0534	1.1974	0.065	0.84	×	P	0.0017	2.2382	0.094	0.90	×	P	2.97	0.046	P
Blennius ocellaris	22	D	7	13.7	10.2	0.1117	1.2885	0.064	0.95	×	P	0.0188	2.264	0.091	0.97	×	P	1.37	0.396	Pn-
Boops boops	75	P	11.2	19.9	15.4	0.1196	1.2863	0.054	0.89	×	P	0.01	2.2408	0.065	0.94	×	P	3.54	0.265	Pn+
Bothus podas	21	D	11.3	17.2	13.5	0.1291	1.1635	0.140	0.78		Li	0.0287	2.0078	0.183	0.86		Li	1.34	0.221	Li
Cepola macrophthalma	84	D	13.2	54.9	31.8	0.1318	0.4876	0.027	0.80	×	P	0.0109	1.4079	0.069	0.84	×	P	0.38	0.419	P*
Chelidonichthys lucernus	15	D	6	21.6	10.7	0.1033	1.2364	0.071	0.96	×	P	0.0147	2.1671	0.072	0.99	×	P	1.47	0.725	Pn-
Chromis chromis	41	RA	8.6	13.3	10.9	0.2582	1.024	0.154	0.53		Li	0.0621	1.7038	0.209	0.63		Li	2.46	0.172	Pn+
Citharus linguatula	95	D	3.9	24.3	14.2	0.135	1.1694	0.02	0.95	×	P	0.0244	2.0817	0.031	0.98	×	P	1.48	0.447	Pn+
Coris julis	49	RA	11.3	18.2	15.8	0.103	1.2629	0.160	0.57		P	0.0181	2.124	0.203	0.70		P	1.79	0.137	Pn+
Dentex dentex	9	BP	11.7	15.3	13.0	0.3396	0.8961	0.320	0.50		Li	0.0167	2.1909	0.265	0.91		E	2.59	0.034	Pn+
Diplodus annularis	72	BP	6.1	17.5	11.1	0.1929	1.117	0.058	0.84	×	P	0.022	2.0722	0.066	0.93		E	2.51	0.086	Pn+
Diplodus vulgaris	38	BP	9	16.7	11.7	0.3501	0.9558	0.103	0.71		Li	0.0452	1.83	0.128	0.85		Li	3.33	0.115	Pn+
Engraulis encrasicolus	83	P	6.7	16.2	11.0	0.2314	0.9106	0.046	0.83	×	P	0.0574	1.3809	0.057	0.90	×	P	2.70	0.397	Pn-
Eutrigla gurnardus	10	D	6.3	14.8	12.8	0.2196	0.9916	0.082	0.95		P	0.0458	1.7016	0.114	0.97	×	P	2.16	0.552	P
Gaidropsarus biscayensis	53	BP	9.1	15.3	12.1	0.0755	1.1428	0.069	0.84	×	P	0.0117	2.0096	0.091	0.91		E	0.97	0.224	Pn+
Gaidropsarus mediterraneus	15	D	8.4	14.5	11.4	0.1055	0.973	0.183	0.69		E	0.0215	1.713	0.190	0.86		P	0.23	0.579	Pn+
Lesueurigobius suerii	47	D	5.8	9.3	7.8	0.0878	1.2393	0.098	0.78	×	P	0.0109	2.4033	0.119	0.90	×	P	0.23	0.558	E
Lophius budegassa	29	D	5.5	38.4	13.2	0.0231	1.5419	0.046	0.98	×	P	0.0043	2.3834	0.076	0.97	×	P	2.41	0.923	Li
Lophius piscatorius	6	D	7.7	12.7	9.6	0.0215	1.558	0.221	0.93	×	P	0.0087	2.1066	0.405	0.87		Li	0.52	0.974	Pn+
Merlangius merlangus	39	BP	14.1	29.1	20.2	0.0788	1.2887	0.067	0.91	×	P	0.0175	2.0713	0.084	0.94		P	1.62	0.592	Pn+
Merluccius merluccius	21	D	11.7	37	20.5	0.0454	1.3958	0.029	0.99	×	P	0.011	2.1227	0.037	0.99	×	P	1.39	0.944	Li
Micromesistius poutassou	47	P	9.2	24	12.2	0.0746	1.2386	0.071	0.87	×	P	0.01	2.089	0.087	0.93		E	1.47	0.352	Pn+
Monochirus hispidus	23	D	9.2	12.8	11.0	1.1012	0.2909	0.336	0.04		Li	0.157	1.2828	0.336	0.41	×	P	1.46	0.249	Pn+
Mullus surmuletus	51	D	9.1	23.1	15.5	0.2367	1.0922	0.030	0.96	×	P	0.0204	2.1134	0.040	0.98	×	P	3.34	0.068	Pn-
Oblada melanura	55	BP	12.6	22.7	18.0	0.178	1.2173	0.054	0.90	×	P	0.0258	2.046	0.080	0.93		L	3.78	0.532	Pn-
Pagellus acarne	52	BP	10.5	19.2	14.7	0.1583	1.2167	0.080	0.82	×	P	0.0247	1.9979	0.100	0.89		E	3.27	0.435	Pn+
Pagellus bogaraveo	64	BP	9.3	22.8	15.1	0.1634	1.2359	0.022	0.98	×	P	0.0197	2.1315	0.027	0.99	×	P	3.39	0.688	Pn+
Pagellus erythrinus	36	BP	8.4	16.4	12.8	0.1196	1.3379	0.132	0.75	×	P	0.0161	2.1964	0.122	0.91		P	3.03	0.179	E
Pagrus pagrus	10	BP	10.2	15.5	12.2	0.2158	1.1379	0.188	0.82		L	0.0165	2.2497	0.146	0.97		Li	3.02	0.227	Pn-

Species	N	H	TL range (cm)		TL_m	TH - TL						TA-TL						Aspect ratio		
			min	max	(cm)	a	B	$SE_{(b)}$	R^2	b≠1	ToR	a	b	$SE_{(b)}$	R^2	b≠2	ToR	A_m	R^2	ToR
Phycis blennoides	27	BP	8.1	37.4	20.4	0.0489	1.2646	0.032	0.99	×	P	0.0077	2.1216	0.033	0.99	×	P	1.04	0.839	P
Pomatomus saltatrix	6	P	13.1	18.5	16.1	0.0765	1.4859	0.110	0.98	×	P	0.0102	2.3803	0.223	0.97		P	1.42	0.970	Pn+
Sardina pilchardus	65	P	7.9	16.7	13.2	0.219	0.9747	0.109	0.56		E	0.0299	1.7494	0.094	0.85	×	P	2.73	0.130	Pn+
Sardinella aurita	76	P	8.9	23.9	18.0	0.3459	0.9054	0.053	0.80	×	P	0.0575	1.5718	0.055	0.92	×	P	4.17	0.136	P
Sarpa salpa	25	BP	11.7	19.5	14.9	0.063	1.5451	0.077	0.95	×	P	0.0044	2.6085	0.092	0.97	×	P	3.28	0.417	Pn-
Sciaena umbra	11	D	12.2	16	14.2	0.0059	2.3463	0.478	0.73	×	P	0.0007	3.4917	0.531	0.83	×	P	1.25	0.461	Pn+
Scomber colias	85	P	8.8	21.7	16.2	0.0582	1.4591	0.075	0.82	×	P	0.009	2.0946	0.063	0.93		P	3.78	0.504	Pn-
Scomber scombrus	79	P	13.4	27.4	21.1	0.0307	1.6515	0.057	0.92	×	P	0.0042	2.3351	0.041	0.98	×	P	4.38	0.829	Pn-
Scorpaena notata	43	D	8.3	17.8	14.2	0.0796	1.4706	0.079	0.89	×	P	0.0141	2.4631	0.080	0.96	×	P	1.60	0.438	E
Scorpaena porcus	69	D	8.2	26.4	14.0	0.0719	1.4471	0.052	0.92	×	P	0.0167	2.3395	0.053	0.97	×	P	1.33	0.669	Li
Scyliorhinus canicula	28	D	29.1	45.1	38.1	0.1343	0.8273	0.089	0.77		E	0.0294	1.6102	0.147	0.82	×	P	0.72	0.034	Pn+
Serranus cabrilla	43	D	9.5	23.1	14.8	0.1357	1.1982	0.063	0.90	×	P	0.0164	2.1648	0.076	0.95	×	P	2.10	0.199	Pn-
Serranus hepatus	61	D	5.7	13.1	9.6	0.2718	0.9589	0.057	0.83		P	0.0556	1.7351	0.067	0.92	×	P	2.01	0.102	Pn-
Serranus scriba	47	D	10.6	23.6	15.9	0.1866	1.0678	0.080	0.80		E	0.033	1.9766	0.106	0.89		P	1.64	0.101	Pn+
Sphyraena sphyraena	29	P	28.3	34.4	31.6	0.9351	0.5448	0.259	0.14	×	P	0.0239	1.7857	0.350	0.49		E	3.32	0.197	Pn-
Spicara maena	83	P	9	20.2	13.7	0.197	1.0563	0.051	0.84		P	0.0253	1.9139	0.059	0.93		P	2.59	0.117	Pn+
Spicara smaris	58	P	7	18.5	11.9	0.1315	1.2354	0.073	0.84	×	P	0.0232	1.93	0.079	0.92		P	0.82	0.744	P
Spondyliosoma cantharus	47	BP	9.7	14	11.6	0.413	0.8187	0.128	0.48		P	0.1106	1.4301	0.138	0.70	×	P	2.58	0.076	Pn-
Symphodus mediterraneus	10	RA	9.8	14.1	12.0	0.8979	0.4808	0.382	0.17		Li	0.1463	1.4759	0.356	0.68		P	1.55	0.154	Pn+
Symphodus tinca	55	RA	11.4	22	15.8	0.2237	0.9867	0.085	0.72		E	0.0238	2.0875	0.097	0.90		E	1.55	0.039	Pn-
Symphurus nigrescens	9	D	6.4	11.1	9.2	0.0356	1.3422	0.126	0.94	×	P	0.0008	2.8561	0.209	0.96	×	P	1.10	0.106	Pn+
Torpedo marmorata	61	D	8.8	36.1	15.9	0.2478	0.9101	0.027	0.95	×	P	0.0385	1.7915	0.037	0.98	×	P	1.73	0.209	Pn-
Trachinus draco	23	D	15	30.5	22.6	0.1022	1.2455	0.105	0.87	×	P	0.0186	2.0407	0.106	0.94		P	2.27	0.265	P
Trachurus mediterraneus	74	P	7	25.8	15.2	0.1215	1.1995	0.047	0.90	×	P	0.0195	1.9248	0.038	0.97	×	P	2.76	0.515	Pn+
Trachurus trachurus	60	P	6.3	23.9	14.9	0.0741	1.3806	0.038	0.97	×	P	0.013	2.0438	0.038	0.98	×	P	2.88	0.837	P
Trisopterus minutus	69	BP	5.7	24.5	12.8	0.0979	1.1917	0.035	0.95	×	P	0.0165	2.069	0.042	0.97		P	1.29	0.556	Pn+
Uranoscopus scaber	55	D	8.7	26.9	14.5	0.2309	1.025	0.040	0.93		L	0.0606	1.86	0.044	0.97	×	P	1.46	0.227	Pn-
Xyrichtys novacula	12	RA	12.3	17.1	14.2	0.0159	2	0.311	0.81	×	P	0.0029	2.8918	0.456	0.8	×	P	1.68	0.808	Pn-

Table 2. Relationships between tail area (TA) and tail height (TH) with total body length (TL) for 61 fishes from the North Aegean Sea, Greece, June 2001- January 2006. N = number

of individuals; H = habitat type; P = pelagic; BP = benthopelagic; D = demersal; RA = reef-associated; SE_b = standard error of slope b; R^2 = coefficient of determination; ToR = type of relationship; E = exponential type; L = logarithmic type; Li = linear type; P = power type; Pn = Polynomial type; A_m = mean tail aspect ratio value. × indicates that b≠2 in the case of TA, and b≠1 in the case of TH. * indicates decreasing trend of the regression line. + indicates cases that polynomial relationship showed a minimum, and – indicates cases that polynomial relationship showed a peak.

The majority of the relationships between A and TL (Table 2) were of the polynomial type (46 out of the 61 species; 75.4%). Out of these 46 species, in 18 species (39.1%) the regression showed a minimum, indicating that A decreases with TL until a certain point and thereafter increases again (e.g., *Trachurus mediterraneus*; Fig. 6), and in the remaining 28 species (60.9%) the regression showed a peak, i.e., A increases with TL up to a maximum and then decreases (i.e., *Engraulis encrasicolus*; Fig. 6). In addition, three other types of relationships were identified for the remaining 15 species: power (8 species; 13.1%); linear (4 species; 6.6%) and exponential (3 species; 4.9%). Among these 15 species, only in the case of *C. macrophthalma* the relationship showed a decreasing trend (Fig. 6).

Finally, the relationship between A_m and TL_m for the 61 species was polynomial (Fig. 7), with a peak at $TL_m \approx 22$-24 cm, indicating that A increases with body length up to a maximum and then decreases.

4. Discussion

The allometric model seems to be the most appropriate for describing morphometrics in fishes (Lleonart et al. 2002) and applies to the vast majority of relationships of morphological characteristics with body length (e.g., Karpouzi & Stergiou 2003, Karachle & Stergiou 2008a, 2010a, b, 2011a). Yet, allometric calculations should not be considered optimally applicable to all metric comparisons, and one must always examine its validity (Peters 1983). Based on the results of the present study, as well as of previous ones (Karachle & Stergiou 2008a, 2010a, b, 2011a), it is also apparent that such relationships might reflect the effect of different factors such as habitat type and feeding habits.

There was a strong effect of body form and shape on W-L relationships and, thus, on b values, since group (I) was comprised of extremely elongated fishes with slim body (i.e., *C. macrophthalma* and *B. belone*), group (II) of elongated, yet more cylindrical body shape (i.e. *S. canicula* and *S. sphyraena*), and group (III) of stream-lined body shape. Furthermore, a values decreased and b values increased from group (I) to group (III), which is in accordance to the widely accepted norm for such relationships in fishes (e.g., Froese 2006). Additionally, based on the results presented here, there was also a strong effect of both feeding habits and habitat type. Nevertheless, the b value of the regressions of species grouped per FTG and habitat type, showed deviation from the widely accepted value b=3 (Froese 2006), a fact that can be mainly attributed to the high dispersion of W-L data values, which resulted from the inclusion of species of different body forms. Thus, these regressions are given only for illustrative purposes.

Indeed, in the case of feeding habits the relationships between FTGs differed significantly, with the exception of that of herbivores. This could be attributed to the fact that only one

Species	N	TL range (cm)	Tail area		Tail height	
			TA - TL	R^2	TH - TL	R^2
Alosa fallax	25	15.0-46.8			TH = 0.1883+0.2782TL	0.94
Anthias anthias	9	12.7-16.6	TA = -24.708+11.635Ln(TL)	0.88		
Apogon imberbis	30	8.0-11.5	TA = 0.4682 $e^{0.2278TL}$	0.70	TH = -0.9077+0.4127TL	0.49
Bothus podas	21	11.3-17.2	TA = -6.2966+0.8666TL	0.90	TH = -0.5264+0.2371	0.83
Chromis chromis	41	8.6-13.3	TA = -3.3347+0.6466TL	0.68	TH = -0.4049 + 0.3124TL	0.58
Dentex dentex	9	11.7-15.3	TA = 0.5339 $e^{0.1651TL}$	0.92	TH = 0.1138 + 0.252TL	0.54
Diplodus annularis	72	6.1-17.5	TA = 0.3658 $e^{0.1918TL}$	0.94		
Diplodus vulgaris	38	9.0-16.7	TA = -4.0126 + 0.6999TL	0.87	TH = -0.018 + 0.3168TL	0.75
Gaidropsarus biscayensis	53	9.1-15.3	TA = 0.2289 $e^{0.1667TL}$	0.91		
Gaidropsarus mediterraneus	15	8.4-14.5			TH = 0.409 $e^{0.0882TL}$	0.70
Lophius piscatorius	6	7.7-12.7	TA = -1.2229 + 0.2378TL	0.91		
Micromesistius poutassou	47	9.2-24.0	TA = 0.3156 $e^{0.143TL}$	0.94		
Monochirus hispidus	23	9.2-12.8			TH = 1.4998 + 0.0667TL	0.04
Oblada melanura	55	12.6-22.7	TA = -43.498 + 18.522Ln(TL)	0.94		
Pagellus acarne	52	10.5-19.2	TA = 0.6787 $e^{0.1387TL}$	0.91		
Pagrus pagrus	10	10.2-15.5	TA = -6.3463 + 0.9042TL	0.97	TH = -6.9748 + 4.2901Ln(TL)	0.86
Sardina pilchardus	65	7.9-16.7			TH = 0.9551 $e^{0.0781TL}$	0.58
Scyliorhinus canicula	28	29.1-45.1			TH = 1.1434 $e^{0.0227TL}$	0.79
Serranus scriba	47	10.6-23.6			TH = 1.1884 $e^{0.0683TL}$	0.80
Sphyraena sphyraena	29	28.3-34.4	TA = 1.8514 $e^{0.0574TL}$	0.50		
Symphodus mediterraneus	10	9.8-14.1			TH = 1.4974 + 0.1232TL	0.17
Symphodus tinca	55	11.4-22.0	TA = 0.9022 $e^{0.1323TL}$	0.90	TH = 1.2383 $e^{0.063TL}$	0.73
Uranoscopus scaber	55	8.7-26.9			TH = 0.0552+0.2438TL	0.93

Table 3. Relationships between tail area (TA) and tail height (TH) with total body length (TL) for 23 fishes from the North Aegean Sea, Greece, June 2001- January 2006, for which power type of the relationship was not statistically significant (for explanation see text). N = number of individuals; R^2 = coefficient of determination.

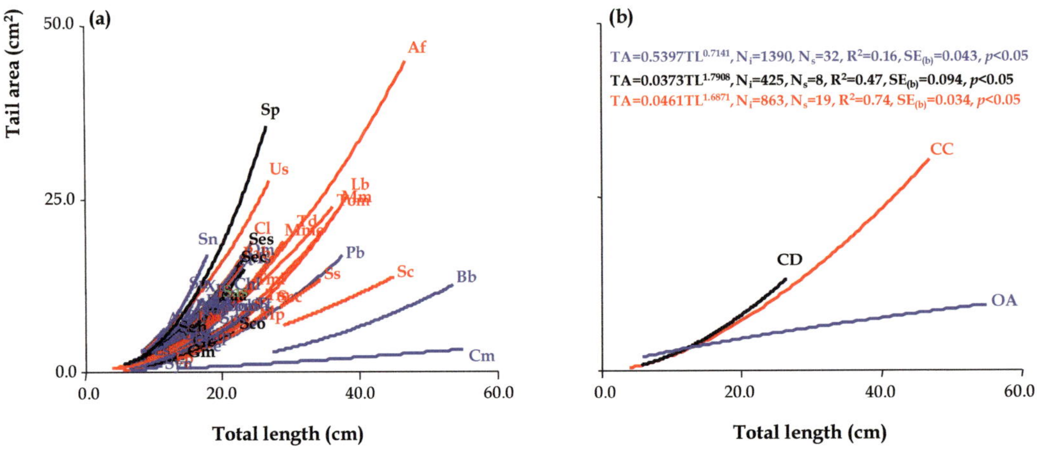

Fig. 4. Regressions between total body length (TL, in cm) and tail area (TA, in cm2) for 61 fish species from the North Aegean Sea, Greece, June 2001- January 2006. Equations are given in Table 2. (a) all regressions of species separately; and (b) regressions of groups of species, according to functional trophic groups based on Stergiou & Karpouzi (2002). Green: herbivores; blue: omnivores with preference to animal material (OA); black: carnivores with preference to decapods and fish (CD); red: carnivores with preference to fish and cephalopods (CC). Af: *Alosa fallax*; Aa: *Anthias anthias*; Ai: *Apogon imberbis*; Al: *Arnoglossus laterna*; Bb: *Belone belone*; Bo: *Blennius ocellaris*; Bob: *Boops boops*; Bp: *Bothus podas*; Cm: *Cepola macrophthalma*; Chl: *Chelidonichthys lucernus*; Cch: *Chromis chromis*; Cl: *Citharus linguatula*; Cj: *Coris julis*; De: *Dentex dentex*; Da: *Diplodus annularis*; Dv: *Diplodus vulgaris*; Ee: *Engraulis encrasicolus*; Eg: *Eutrigla gurnardus*; Gb: *Gaidropsarus biscayensis*; Gm: *Gaidropsarus mediterraneus*; Ls: *Lesueurigobius suerii*; Lb: *Lophius budegassa*; Lp: *Lophius piscatorius*; Mme: *Merlangius merlangus*; Mm: *Merluccius merluccius*; Mp: *Micromesistius poutassou*; Mh: *Monochirus hispidus*; Ms: *Mullus surmuletus*; Om: *Oblada melanura*; Paa: *Pagellus acarne*; Pab: *Pagellus bogaraveo*; Pae: *Pagellus erythrinus*; Pp: *Pagrus pagrus*; Pb: *Phycis blennoides*; Ps: *Pomatomus saltatrix*; Sap: *Sardina pilchardus*; Sa: *Sardinella aurita*; Sas: *Sarpa salpa*; Su: *Sciaena umbra*; Sco: *Scomber colias*; Scs: *Scomber scombrus*; Sn: *Scorpaena notata*; Sp: *Scorpaena porcus*; Sc: *Scyliorhinus canicula*; Sec: *Serranus cabrilla*; Seh: *Serranus hepatus*; Ses: *Serranus scriba*; Ss: *Sphyraena sphyraena*; Spm: *Spicara maena*; Sps: *Spicara smaris*; Spc: *Spondyliosoma cantharus*; St: *Symphodus tinca*; Syn: *Symphurus nigrescens*; Tom: *Torpedo marmorata*; Td: *Trachinus draco*; Tm: *Trachurus mediterraneus*; Tt: *Trachurus trachurus*; Tmi: *Trisopterus minutus*; Us: *Uranoscopus scaber*; Xn: *Xyrichtys novacula.*

herbivore species, namely *Sarpa salpa*, was included in the dataset, with a low number of individuals. Thus, this must be verified by including more herbivorous species. Additionally, based on the results presented here, for a given length carnivores are more robust than omnivores. The effect of diet could be anticipated, since, with food, organisms attain the necessary energy and nutrients for somatic growth and reproduction. Carnivorous feeding is considered as more profitable in terms of energy, whereas herbivorous diets or inclusion of plants in the daily "menu" (such as in the case of omnivores) requires larger quantities of food (e.g., Gerking 1994, Wootton 1998) or morphologic adaptations (e.g., longer intestines: Wootton 1998, Pennisi 2005, Karachle & Stergiou 2010b) to meet with energetic demands. When habitat type was examined, there was also a significant difference in the weight-length relationships. From an ecological point of view, body form and habitat of any given species are strongly related. In general, pelagic and benthopelagic species are characterised by a more stream-lined body shape, reef-associated species are more roundish and demersal species seem to have more or less compressed (both dorsoventrally and laterally) bodies. This ecological adaptation in body form is also reflected in the weight-length relationships, since the only case where no difference was identified in the weight-length regressions was between pelagic and benthopelagic species; in all the remaining combinations the differences in weight-length relationships were highly significant.

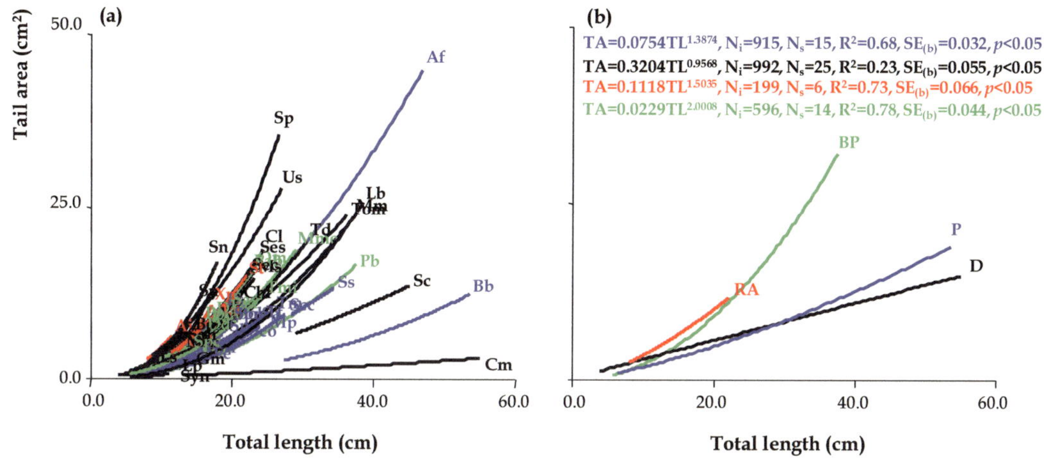

Fig. 5. Regressions between total body length (TL, in cm) and tail area (TA, in cm^2) for 61 fish species from the North Aegean Sea, Greece, June 2001- January 2006. Equations are given in Table 2. (a) all regressions of species separately; and (b) regressions of groups of species, according to habitat type (from FishBase, www.fishbase.org: Froese & Pauly 2011). Blue: pelagic (P); green: benthopelagic (BP); black: demersal (D); red: reef-associated (RA). Abbreviations of species names are given in figure 4.

It has been previously shown (Karpouzi & Stergiou 2003, Karachle & Stergiou 2011a) that for the same body length omnivorous fishes tend to have smaller mouth area than carnivorous ones. The larger mouths of carnivores could be attributed mainly to: (a) adaptations of the structural capacity in order to meet with increasing energetic demands (Galis et al. 1994) and (b) more effective handling and consumption of prey with large size (Scharf et al. 2000, Pauly et al. 2001). Indeed, according to the optimal foraging theory (Gerking 1994), carnivorous fishes that mainly feed with prey of high motility (e.g., other fishes) need to consume higher amounts of food or food of larger size in fewer feeding attempts, a fact that can be achieved by larger mouth gape and other adaptations (e.g., vision acuity, fast swimming and effective digestion; Keast & Webb 1966, Kaiser & Hughes 1993, Juanes 1994, Juanes & Conover 1994, Hart 1997, Wootton 1998, Fordham & Trippel 1999). On the other hand, there was no clear effect of habitat on mouth area. For example, in the category of reef-associated species are included species with large differences in mouth size; *Apogon imberbis* and *Anthias anthias* are two such species, which prey on fishes and benthic crustaceans (Karachle & Stergiou 2010c, www.fishbase.org), with rather big mouth gapes, whereas *Coris julis* and *Symphodus tinca* (which prey upon worms, bivalves, gastopods and small crustaceans, such as amphipods (Karachle & Stergiou 2010c, www.fishbase.org) are two species with small mouths, yet strong dentition. Likewise, pelagic species include a wide range of predators: from small pelagic filter feeders, such as sardines and anchovies, which prey on small-sized zooplankton (copepods) (www.fishbase.org) to apex predators such as tunas, which feed on fishes (www.fishbase.org). The same is also true for the remaining two categories of fishes (demersal: *Lesueurigobius suerii* and *Blennius ocellaris* that prey on small crustaceans and molluscs (Karachle & Stergiou 2010c), with small mouths, and flatfishes, such as *Arnoglossus laterna* and *Citharus linguatula*, that prey on fishes (Karachle & Stergiou 2011b) with large mouth gape; benthopelagic: *Diplodus annularis* and *Oblada melanura*, that mainly feed on worms, molluscs and small crustaceans, with small mouths, and large-mouthed species as *Gadus morhua* and *Merlangius merlangus*, that prey on fishes (www.fishbase.org).

With respect to intestine length, there was a strong grouping of species according to their feeding habits: species that fed exclusively on plants and those which included large amounts of vegetable material in their diet (omnivores with preference to plants) formed a group that clearly separated from carnivorous species (omnivores with preference to animal material, carnivores with preference to fish and decapods and carnivores with preference to fish and cephalopods). Additionally, for the same body length, species of the first group had remarkably longer intestines than species in the second one. The above differences mainly result from the fact that plant material is more resistant to digestion, and hence longer intestines are required in order adequate amounts of nutrients and energy to be absorbed (e.g., Wootton 1998, Pennisi 2005). The effect of habitat type was also clear, yet no difference was observed between the pelagic and benthopelagic species studied here, a fact also observed in the case of weight-length relationships, and can be attributed to the fact that intestine growth, form and shape is strongly affected by the general body form, which in turn, as mentioned above, is related to habitat type (Verigina 1991, Karachle & Stergiou 2010b).

The effect of both feeding habits and habitat type on tail morphometrics was also strong, with TA for the same body length increasing faster in carnivorous than in omnivorous species, and for benthopelagic than pelagic and demersal species. This fact can be attributed both to the differentiation of the general body form of fishes with habitats, as mentioned above, and to the importance of tail shape and area to the acquisition of food (Keast & Webb 1966, Ward-Campbell & Beamish 2005).

Despite the extensive search for relative literature on tail characteristics, and especially on relationships linking TA and TH to body length, no such information was found. The only available information is restricted to estimates of A, and the majority of such estimates are from photographs and/or fish drawings (www.fishbase.org). Additionally, there are differences in A estimates for the same species, a fact that has been attributed to one or a combination of the following parameters (García & Duarte 2002): (a) the method used for the estimation of TH and TA, (b) the type of picture (i.e., photograph or drawing) used for A estimation and (c) the disposition of tail. According to Palomares & Pauly (1998) the most appropriate way of acquiring more accurate A estimates is a disposition of tail resembling that of swimming position and estimating TA to the point where caudal peduncle was the lowest height. The prevalence of the polynomial type in the relationships between A and TL adds a further factor responsible for such differences, notably the body length of the specimen used for A estimation. Hence, when A values are needed for the estimation of Q/B ratio, the body length of the species in question should be taken into consideration as well. Nevertheless, this needs further investigation, since there were 15 species that did not show polynomial type of relationship. Among these 15 species, only in the case of *C. macrophthalma* the relationship showed a decreasing trend. This could be attributed to the fact that tail shape in this species differs with sex (Stergiou 1991): in males the central spines of the caudal fin are rather elongated and therefore tail total area is larger than that in females, and hence A is lower. It must be stressed that the length range of the individuals of each sex used in the present study differed, with males being generally lengthier than females (males: 18.2-54.9 cm; females: 13.2-47.6 cm).

Additionally, the relationship between A_m and TL_m of the examined 61 species was also polynomial, showing a maximum at $TL_m \approx 22$-24 cm. The decline after this threshold of TL should be attributed to the fact that species which are located to the right of the peak are those with a rather elongated body form (i.e., *B. belone, C. macrophthalma, S. canicula* and *S. sphyraena*). This agrees with Pauly (1989b) who maintains that body depth ratio (i.e., the ratio between body length and maximum body depth) is positively related to food consumption, which, in turn, is positively related to A. Hence, fishes with a rather elongated body form, and therefore low body depth ratios, should be expected, for the same TL, to have lower A values than species with a streamlined or diamond-shaped body.

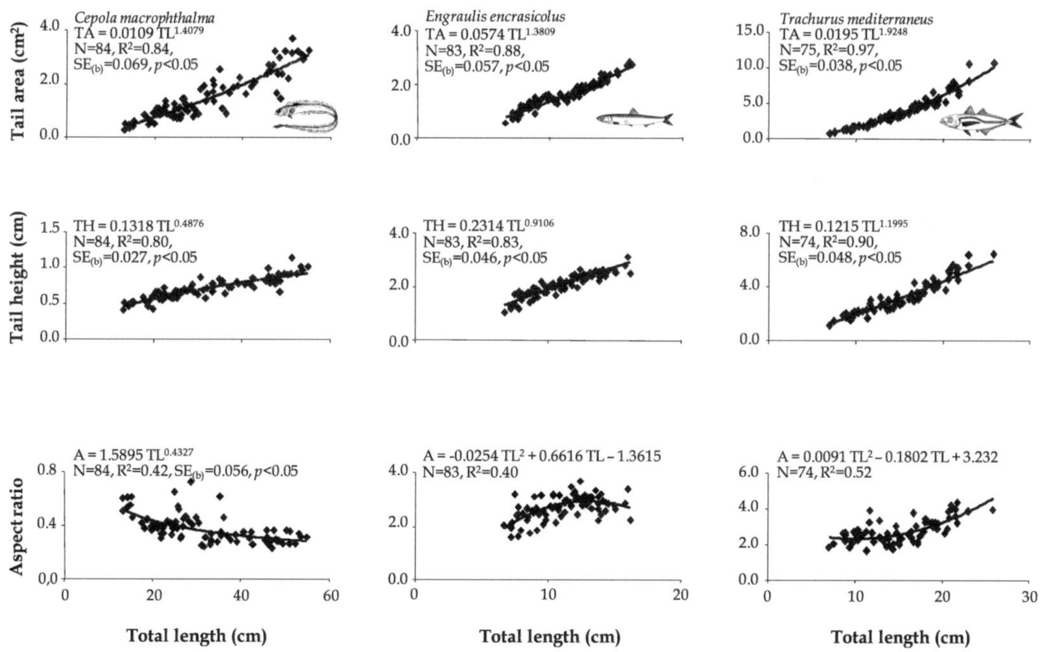

Fig. 6. Relationships between tail area (TA; top), tail height (TH; center) and tail aspect ratio (A; bottom) and total body length (TL) for 3 fish species, from the North Aegean Sea, Greece, June 2001- January 2006. Fish drawings from FishBase (www.fishbase.org: Froese & Pauly 2011).

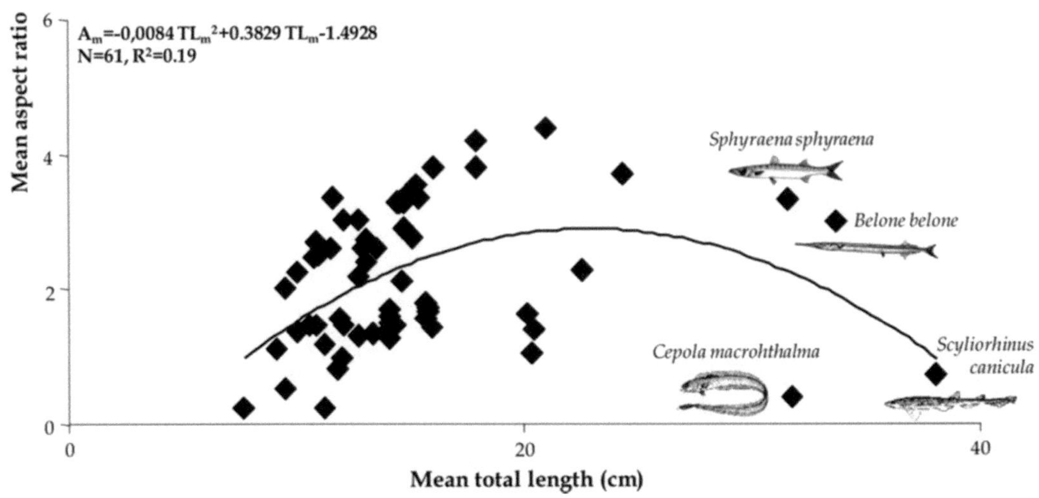

Fig. 7. Relationship between the mean values of tail aspect ratio (A_m) and mean total body length (TL_m) for 61 fish species, from the North Aegean Sea, Greece, June 2001- January 2006. Fish drawings from FishBase (www.fishbase.org: Froese & Pauly 2011).

5. Conclusions

1. The allometric model is the most appropriate in describing morphometric relationships in fishes, yet its validity should not be taken for granted.
2. There is a strong effect of feeding habits on the way the morphological characteristics presented here (weight, mouth, intestine and tail) change with body length.
3. Habitat type was found to affect the way that weight, intestine and tail change with body length. Yet, this is not true of mouth area.
4. Tail aspect ratio shows, in the majority of cases, a polynomial type of relationship with total body length. The only case where there was a decrease in tail aspect ratio with size was for *Cepola macrophthalma,* in which there is a strong a sexual dimorphism in tail size/shape.
5. Mean tail aspect ratio is related to total body length with a polynomial type of model, showing a peak at a mean TL of ≈22-24 cm.
6. The polynomial type of model found between 46 out of the 61 species (75.4%) indicates that when estimating Q/B ratios, the body length of the individuals should also be taken into account.

6. References

Al-Hussaini, A.H. (1947). The Feeding Habits and the Morphology of the Alimentary Tract of Some Teleosts Living in the Neighbourhood of the Marine Biological Station, Ghardaqa, Red Sea. *Publications of the marine Biological Station, Ghardaqa (Red Sea),* Vol.5, pp. 1-61, ISSN 0370-0534

Angelini, R. & Agostinho, A.A. (2005). Parameter Estimates for Fishes of the Upper Paraná River Floodplain and Itaipu Reservoir (Brazil). *NAGA, World Fish Center Newsletter,* Vol.28, No.1 & 2, pp. 53-57, ISSN 0115 4575

Binohlan, C. & Pauly, D. (2000). The length-weight table, In: *Fishbase 2000: concepts, design and data sources,* Froese R. & D. Pauly, (Eds), 121-123, ICLARM, ISBN 971-8709-99-1, Manila, Philippines

Chalkia, V.V. & Bobori, D.C. (2006). Length-Length and Length-Mouth Dimensions Relationships of Freshwater Fishes of Northern Greece. *Proceedings of the 10th International Congress on the Zoogeography and Ecology of Greece and Adjacent Regions* pp. 123. Patra, Greece, June 26-30, 2006

Chivers, D.J. & Hladik, C.M. (1980). Morphology of the Gastrointestinal Tract in Primates: Comparisons with other Mammals in Relation to Diet. *Journal of Morphology,* Vol.166, pp. 337-386, ISSN 1097-4687

Fordham, S.E. & Trippel, E.A. (1999). Feeding Behaviour of Cod (*Gadus morhua*) in Relation to Spawning. *Journal of Applied Ichthyology,* Vol.15, pp. 1-9, ISSN 1439-0426

Froese, R., Pauly, D. (Eds.) (2011). FishBase. World Wide Web electronic publication. URL: www.fishbase.org http//:www.fishbase.org [version 08/2011]

Galis, F., Terlouw, A. & Osse, J.W.M. (1994). The Relation between Morphology and Behaviour During Ontogenetic and Evolutionary Changes. *Journal of Fish Biology,* Vol.45, pp. 13-26, ISSN 1095-8649

García, C.B. & Duarte, L.O. (2002).Consumption to Biomass (Q/B) Ratio and Estimates of Q/B-Predictor Parameters for Caribbean Fishes. *NAGA, The ICLARM Quarterly,* Vol.25, No.2, pp. 19-31, ISSN 0116-290X

Gerking, S.D. (1994). Feeding Ecology of Fish. Academic Press, ISBN 0-12-280780-4, San Diego, USA

Hart, P.J.B. (1997). Foraging Tactics, In: *Behavioural Ecology of Teleost Fishes,* J.G.J. Godin, (Ed.), 104–133, Oxford University Press, ISBN 0-19-850503-5, New York, USA

Juanes, F. & Conover, D.O. (1994). Piscivory and Prey Size Selection in Young-of-the-Year Bluefish: Predator Preference or Size-Dependent Capture Success? *Marine Ecology Progress Series,* Vol.114, pp. 59-69, ISSN 0171-8630

Juanes, F. (1994). What Determines Prey Size Selectivity in Piscivorous Fishes? In: *Theory and application in fish feeding ecology,* Stouder, D.J., Fresh, K.L. & Feller, R.J., (eds), 78-100, Belle W. Baruch Library in Marine Sciences, no. 18, ISBN 1570030138, University of South Carolina Press, Columbia, SC, USA

Kaiser, M.J. & Hughes, R.N. (1993). Factors Affecting the Behavioural Mechanisms of Diet Selection in Fishes. *Marine Behaviour and Physiology,* Vol.23, pp. 105-118, ISSN 0091-181X

Kapoor, B.G., Smit, H. & Verighina, I.A. (1975). The Alimentary Canal and Digestion in Teleosts. *Advances in Marine Biology,* Vol.13, 109-239, ISSN 0065-2881

Karachle, P.K. & Stergiou, K.I. (2004). Preliminary Results on Relationships between Tail Area and Total Body Length for Four Fish Species. *Rapports de la Commission International de la Mer Méditerranée,* Vol.37, pp. 376

Karachle, P.K. & Stergiou, K.I. (2005). Morphometric Relationships in Fishes. *Proceedings of the 3rd FishBase Mini Symposium "Fish and More",* pp. 45-47, Thessaloniki, Greece, August 31, 2005

Karachle, P.K. & Stergiou, K.I. (2008a). Tail Shape of Various Fishes. *Proceedings of the 30th Scientific Conference of the Hellenic Association for Biological Sciences*, pp. 188-189, Thessaloniki, Greece, May 22-24, 2008, ISSN 1109-4885

Karachle, P.K. & Stergiou, K.I. (2008b). Length-Length and Length-Weight Relationships of Several Fish Species from the North Aegean Sea (Greece). *Journal of Biological Research*, Vol.10, pp. 149-157, ISSN 1790-045X

Karachle, P.K. & Stergiou, K.I. (2008c). The Effect of Season and Sex on Trophic Levels of Marine Fishes. *Journal of Fish Biology*, Vol.72, pp. 1463-1487, ISSN 1095-8649

Karachle, P.K. & Stergiou, K.I. (2010a). Intestine Morphometrics: a Compilation and Analysis of Bibliographic Data. *Acta Ichthyologica et Piscatoria*, Vol.40, No.1, pp. 45-54, ISSN 0137-1592

Karachle, P.K. & Stergiou, K.I. (2010b). Gut Length for Several Marine Fishes: Relationships with Body Length and Trophic Implications. *Marine Biodiversity Records*, 3: e106, pp. 1-10, ISSN 1755-2672

Karachle, P.K. & Stergiou, K.I. (2010c). Food and Feeding Habits for Eight Fish Species from the North Aegean Sea. *Proceedings of the 14thPanhellenic Symposium of Ichthyologists*, pp. 7-10, Piraeus, Greece, May 6-9, 2010

Karachle, P.K. & Stergiou, K.I. (2011a). Mouth Allometry and Feeding Habits in Fishes. *Acta Ichthyologica et Piscatoria*, Vol.41, No.4, pp. 265-275, ISSN 0137-1592

Karachle, P.K. & Stergiou, K.I. (2011b). Feeding and Ecomorphology for Seven Flatfishes in the N-NW Aegean Sea (Greece). *African Journal of Marine Science*, Vol.33, No.1, pp. 67-78, ISSN 1814-232X

Karpouzi, V.S. & Stergiou, K.I. (2003). The Relationships between Mouth Size and Shape and Body Length for 18 Species of Marine Fishes and Their Trophic Implications. *Journal of Fish Biology*, Vol.62, pp. 1353-1365, ISSN 1095-8649

Keast, A. & Webb, D. (1966). Mouth and Body Form Relative to Feeding Ecology in the Fish Fauna of a Small Lake, Lake Opinicon, Ontario. *Journal of Fish Research Board of Canada*, Vol.23, No.12, pp. 1845-1874, ISSN 0015-296X

Kramer, D.L. & Bryant, M.J. (1995a). Intestine Length in the Fishes of a Tropical Stream: 1. Ontogenetic Allometry. *Environmental Biology of Fishes*, Vol.42, pp. 115-127, ISSN 0378-1909

Kramer, D.L. & Bryant, M.J. (1995b). Intestine Length in the Fishes of a Tropical Stream: 2. Relationships to Diet – the Long and Short of a Convoluted Issue. *Environmental Biology of Fishes*, Vol.42, pp. 129-141, ISSN 0378-1909

Lleonart, J., Salat, J. & Torres, G.J. (2000). Removing Allometric Effects of Body Size in Morphological Analysis. *Journal of Theoretical Biology*, Vol.205, pp. 85-93, ISSN 0022-5193

Motta, P.J., Norton, S.F. & Luczkovich, J.J. (1995). Perspectives on the Ecomorphology of Bony Fishes. *Environmental Biology of Fishes*, Vol.44, pp. 11-20, ISSN 0378-1909

O'Grady, S.P., Morando, M., Avila, L. & Dearing, M.D. (2005). Correlating Diet and Digestive Tract Specialization: Examples from the Lizard Family Liolaemidae. *Zoology*, Vol.108, pp. 201-210, ISSN 0944-2006

Palomares, M.L. & Pauly, D. (1989). A Multiple Regression Model for Predicting the Food Consumption of Marine Fish Populations. *Australian Journal of Marine and Freshwater Research*, Vol.40, pp. 259-273, ISSN 0067-1940

Palomares, M.L. & Pauly, D. (1998). Predicting Food Consumption of Fish Populations as Functions of Mortality, Food Type, Morphometrics, Temperature and Salinity. *Australian Journal of Marine and Freshwater Research*, Vol.49, pp. 447-453, ISSN 0067-1940

Pauly, D. (1989a). A Simple Index of Metabolic Level in Fishes. *Fishbyte*, Vol.7, No.1, pp. 22, ISSN 0116-0079

Pauly, D. (1989b). Food Consumption by Tropical and Temperate Fish Populations: Some Generalizations. *Journal of Fish Biology*, Vol.35, No. Supplement A, pp. 11-20, ISSN 1095-8649

Pauly, D. (1993). Fishbyte Section. Editorial. *NAGA, The ICLARM Quarterly*, Vol.16, pp26, ISSN 0116-290X

Pauly, D., Palomares, M.L., Froese, R., Sa-a, P., Vakily, M., Preikshot, D. & Wallace, S. (2001). Fishing Down Canadian Aquatic Food Webs. *Canadian Journal of Fisheries and Aquatic Sciences*, Vol.58, pp. 51–62, ISSN 0706-652X

Pennisi, E. (2005). The dynamic gut. What's eating you? *Science*, Vol.307, pp. 1896-1899, ISSN 0036-8075

Peters, R.H. (1983). The ecological implications of body size. Cambridge University Press, ISBN 0-521-28886-X, New York, USA

Petrakis, G. & Stergiou, K.I. (1995). Weight-Length Relationships for 33 Fish Species in Greek Waters. *Fisheries Research*, Vol.21, pp. 465-469, ISSN 0165-7836

Ribble, D.O. & Smith, M.H. (1983). Relative Intestine Length and Feeding Ecology of Freshwater Fishes. *Growth*, Vol.47, pp. 292-300.

Ricklefs, R.E. (1996). Morphometry of the Digestive Tracts of Some Passerine Birds. *Condor*, Vol.98, pp. 279-292, ISSN 0010-5422

Scharf, F.S., Juanes, F. & Rountree, R.A. (2000). Predator Size–Prey Size Relationships of Marine Fish Predators: Interspecific Variation and Effects of Ontogeny and Body Size on Trophic Niche Breadth. *Marine Ecology Progress Series*, Vol.208, pp. 229-248, ISSN 0171-8630

Stergiou, K.I. & Karpouzi, V.S. (2002). Feeding Habits and Trophic Levels of Mediterranean Fish. *Reviews in Fish Biology and Fisheries*, Vol.11, pp. 217-254, ISSN 0960-3166

Stergiou, K.I. (1991). Biology, ecology and dynamics of *Cepola macrophthalma* (L., 1758) (Pisces Cepolidae) in the Euboikos and Pagasitikos Gulfs. PhD Thesis, Department of Biology, Aristotle University of Thessaloniki, Greece, 221 pp.

Verigina, I.A. (1991). Basic Adaptations of the Digestive System in Bony Fishes as a Function of Diet. *Journal of Ichthyology*, Vol.31, No.2, pp. 8-20, ISSN 0032-9452

Wainwright, P.C. & Richard, B.A. (1995). Predicting Patterns of Prey Use from Morphology of Fishes. *Environmental Biology of Fishes*, Vol.44, pp. 97-113, ISSN 0378-1909

Ward-Campbell, B.M.S. & Beamish, F.W.H. (2005). Ontogenetic Changes in Morphology and Diet in the Snakehead, *Channa limbata*, a Predatory Fish in Western Thailand. *Environmental Biology of Fishes*, Vol.72, pp. 251-257, ISSN 0378-1909

Wilcox, C.D., Dove, S.B., McDavid, W.D. & Greer, D.B. (1997). UTHSCSA Image Tool User Manual. University of Texas Health Science Center, San Antonio, Texas.

Wootton, R.J. (1998). Ecology of teleost fishes. 2nd Edition. Kluwer Academic Publishers, ISBN 0-412-64200-X, Fish and Fisheries Series 24, London

Zar, J.H. (1999). Biostatistical analysis. 4th Edition. Prentice Hall, ISBN 0-13-081542-X, New Jersey, USA.

5

Morphometry Applied to the Study of Morphological Plasticity During Vertebrate Development

Christina Wahl
Wells College, Aurora, NY
USA

1. Introduction

Embryos and young growing animals do not develop in isolation. The lake shiner, *Notropis atherinoides,* and the blue-gill sunfish, *Lepomis incisor,* produce more body segments when raised in cool water. Chickens raised in constant light have flattened corneas, making them abnormally hyperopic, or "far-sighted". Cichlids provided with different types of diets during early growth develop different jaw morphologies as adults. Craniofacial proportions are different in children born with fetal alcohol syndrome, and those born to women who smoke tend to be underweight compared to the average. These examples demonstrate the plasticity of shape and size that is possible during ontogeny as a result of environmental conditions, and all produce permanent effects on the adult phenotype. In this chapter, I will describe different forms of vertebrate developmental phenotypes and phenotypic plasticity, with a brief review of the relevant biology, and then I will present some preliminary approaches to morphometric quantification of these phenomenae.

The *metamorphosis* of a unique embryonic, or larval, body type into the definitive adult body form of the species (such as seen in fishes) involves dramatic, permanent phenotypic change, whereas *regeneration,* a property that the embryos of many species possess to a remarkable degree, is a form of phenotypic plasticity that effects embryonic repairs. Metamorphosis is highly refined among the invertebrates, in particular among insects, however some vertebrates (fishes) exhibit metamorphosis too, and quite spectacularly (see Figure 1). Among the mesopelagic Stomiiform fishes, larval craniofacial features include elaborate larval eyestalks and elongated, dorso-ventrally flattened skulls, which transform during metamorphosis into a more typical face...eyes seated within orbital sockets instead of at the ends of eye stalks…and increased skull depth. Some species of fishes actually shrink in size as well as change their shape during metamorphosis. For instance, the leptocephalus larvae of anadromous eels is significantly larger than the adult of the species.

Embryonic regeneration is spectacular…it is the ability to achieve scarless reconstruction of injured body parts…and can extend from replacement of missing limbs to functional repair of enucleated eyes, as noted among amphibians in the order Urodela. The ability to regenerate is more robust among embryos and among the young, although some vertebrates, such as the urodelans, retain vigorous regenerative capacity throughout life.

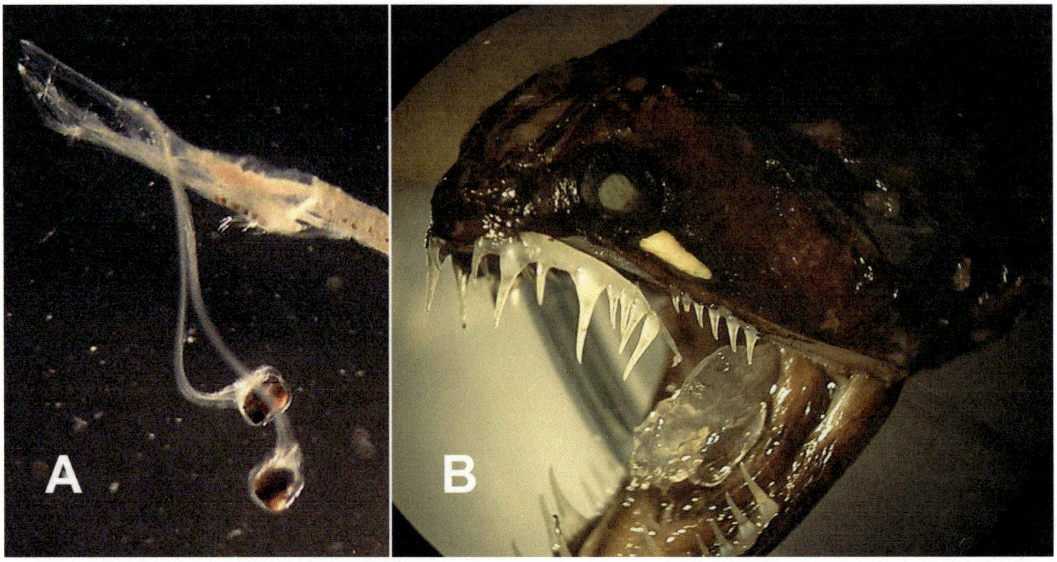

Fig. 1. Larval dragonfish *Idiacanthus atlanticus* (a) standard length (SL) 3.5 cm, and adult (b) SL 13 cm. Craniofacial metamorphosis is pronounced, especially around the periocular area. Larval eye stalks are up to 1/3 of total body length, but are absent in the adult. The adult specimen shown here has partially ingested a jelly, still visible in its throat (Specimens photographed by C. Wahl at CSIRO, Hobart, and at the Australian Museum, Sydney, Australia).

Although gene expression is at the heart of development and determines the basic "bauplan", specific details of morphology such as size, shape, numbers of body segments, and even sex can be strongly influenced by the embryonic and early life environment, principally through action on signaling pathways or through gene regulation. In recent years, new attention has been paid to the ways in which developmental mechanisms are able to produce specific phenotypic solutions to environmental variables (Müller, 2011).

Flexibility to adapt the body to local conditions during growth confers on the embryo, larva, or juvenile an opportunity to fine-tune certain aspects of anatomy and physiology and may increase fitness as the individual reaches adulthood. Epigenetic influence on development may prove to be not only common, but in many cases, critical to adaptive evolutionary change.

Whereas the healing of amputated limbs or enucleated eyes in salamanders can result in slightly smaller yet functional replacements (Wahl, 1985), among embryos perfectly scaled repairs are possible (Wahl and Noden, 2001). The mechanisms active during embryonic regeneration must co-exist in the same body with temporally disparate ontogenetic activity....because one part of the body is being completely re-built while the rest is already further along the road towards adulthood.

Development is defined, for the purposes of this chapter, as the process of transformation of haplo-diploid organisms from zygote to sexual maturity. The "*embryonic period*" is defined as the time during development when the organism is unable to live without either a yolk sac or a placenta. The term "*fetal*" is used most often by medical professionals in reference to the latter period of human gestation, but since the exact developmental interval it refers to is poorly defined and does not apply to other vertebrates, the term is not useful to the basic science of developmental biology.

Morphometry, i.e., methods used to quantify body forms, can be usefully applied to many questions in development, ecology, and physiology. Morphometric assessment of developing organisms offers valuable insights into the consequences of both natural and pathological environmental variables, and usefully informs analysis of epigenetic influence on gene expression patterns. This chapter introduces the different types of morphological variation among vertebrate embryos, and discusses some morphometric assessment techniques. As a colleague has pointed out, "It may be worth noting in passing that shape is a qualitatively discrete character; it is only our insufficient description of it that forces us to rely on continuous measures." (McCune, 1981) . Quantitative measurements allow us to evaluate incremental changes in shape and size, both without and within bodies and organs.

2. Adaptive and maladaptive morphological plasticity

Growth is either uniform or allometric (disproportionate), and each type occurs naturally both within and among species. An example of uniform growth is bilateral symmetry. However, the developmental bauplan also necessitates allometry; limbs cannot grow properly without the prior appearance of the nervous and circulatory systems. The vertebrate head is usually disproportionately large throughout the embryonic period in order to prioritize development of the brain, eyes, and mouthparts. Allometry as a manifestation of morphological integration occurring during development has been studied for many years (Klingenberg, 2008).

Environmental stressors alter growth patterns, and this is important to recognize and quantify morphometrically. It may also be important to distinguish between stressors that affect uniform growth versus those that influence allometric growth, since these have different implications to both short and long term fitness and viability.

2.1 Uniform growth

As D'Arcy Thompson pointed out in 1917, with respect to biological systems, as an organism increases in size, the forces in action within its systems vary. For instance, some physical features scale as functions of the mass, others scale with volume. While the "dimensions" may remain the same in our equations of equilibrium, the relative values alter with scale (Bonner, 1969). The consequences of this "principle of similitude", first described by Galileo, has implications at every level of the developing body. Thus, gravity is of consequence to the whole animal only after it reaches a certain size, putting constraints on the maximum size attainable--but gravity is not a significant force to the neurulating embryo, where other properties such as diffusion gradients and turgor pressure are more important. The young embryo relies on direct diffusion of oxygen to the tissues prior to the development of its circulation, and makes use of turgor pressure to expand the brain and create various body folds. Such forces as viscosity and surface tension can have enormous influence on body form during this period.

Aside from genetic malfunctions, variations of developmentally significant environmental parameters produce asymmetries of uniform scaling, known as "fluctuating asymmetry" (Van Valen, 1962). Because bilateral asymmetries originate from random perturbations of developmental processes (Klingenberg, 2003), such asymmetries must arise within the developmental pathways themselves. Possibly because symmetry is a measure of developmental stability, among vertebrates it has been shown that bilaterally symmetrical individuals are more attractive than asymmetrical individuals to members of the opposite sex (Etcoff, 2000).

What is the scholarly interest in attending to variations of uniform scaling among developing organisms? In the evolutionary context, it is because interactions between developmental pathways have significant effects on the phenotypic outcome and stabilizing selection should limit variation. However, adaptive plasticity to environmental parameters is also a survival strategy and is the mechanism by which the choice is made to mature at smaller or larger body sizes, a response to limited resources known as the "thrifty phenotype".

"The thrifty phenotype is the consequence of three different adaptive processes - niche construction, maternal effects, and developmental plasticity… The three processes also operate at different paces... In contemporary populations, the sensitivity of an offspring's development to maternal phenotype exposes the offspring to adverse effects, through four distinct pathways. The offspring may be exposed to (1) poor maternal metabolic control (e.g. gestational diabetes), (2) maternally derived toxins (e.g. maternal smoking), or (3) low maternal social status (e.g. small size)." (Wells, 2007).

During nutritional dearth, an individual may complete development at a smaller body size or mass than when the nutritional status is excellent. Smaller, metabolically less active individuals produced on limited nutrition exhibit this "thrifty phenotype", demonstrated in several species, including rats (Buresova et al., 2006). Understanding how the thrifty phenotype is generated, and what the long-term consequences of such a phenotype might be, is currently of great interest due to the rising incidence of obesity and type 2 diabetes among western civilizations (Wells, 2007). These "diseases of the wealthy" are regarded by some as a maladaptive response to calorie-rich but nutritionally inadequate prenatal diets, where offspring, like their mothers, continue to consume more calories than their "thrifty" metabolism is equipped to burn.

2.2 Allometric growth

Normal developmental patterns of allometry and variations due to selection pressures are a topic of long-standing interest to evolutionary biologists, spawning the field of "evo-devo". Some growth patterns may be a "normal" consequence of the immediate environment, for instance, differences correlated with temperature include shorter limbs among endotherms at higher latitudes (Allen's Rule) for which a possible mechanism has recently been discovered… mice raised at lower temperatures have shorter limbs than littermates raised at higher temperatures (Serrat et al., 2008). Another example is that more body segments differentiate among fish of a given species developing in cool water, than are found in conspecifics raised in warmer water. Further study of this meristic and others correlated with temperature might reveal whether Bergman's Rule (the reduction of surface-to-volume ratio with reduced environmental temperature) or Allen's Rule (shorter limbs at lower environmental temperatures) are manifestations of epigenetic responses by the developing animal. There are also pragmatic reasons to count fish vertebrae or make other measurements of animals from different climates. This information could be of assistance, along with DNA fingerprinting, in identifying fishing violations, or in determining where a particular animal "grew up".

Thompson defined allometry as "the study of size and its consequences" (Bonner, 1969), and there are physiological consequences of size and scale within the developing organism that are related to constraints of integration of developing body systems. Many allometric relationships scale as the power function $y = bx^a$, where x and y are the two traits being compared (for instance, height and weight), b is the y-intercept, and α is the slope. However, any linear relationship with a y-intercept greater than 0 can describe allometric growth. For a detailed discussion of the use, and misuse, of mathematical relationships that describe allometric growth the reader is referred to many excellent reviews on the subject, for instance, (Gould, 1966).

The response of developing organisms to local environment with phenotypic adaptation has been termed "epigenetic innovation".

"The fact that perturbations of general developmental parameters, such as blastema size, timing of processes, inductive interactions, or cell division rates, could yield very specific morphogenetic results that (mimic) patterns observed in natural change (is) a strong indication that the rules of ontogenetic development (have) an impact on the process of evolutionary variation." (Müller, 2011).

This idea has profound implications when one considers how developmental pathologies arise as a result of conditions such as hypoxia, hypertension, hyperglycemia, and the like. If embryos are capable of "epigenetic innovation" in a single generation, then environment.... influenced, among amniotes, by such factors as maternal diet and behavior....will affect both the morphological and physiological phenotype of the offspring, just as external environmental parameters like temperature directly affect the development of poikilotherms.

Examples of abnormal variations in the embryonic environment include placental insufficiencies, fungal/viral infections, and teratogens. The effects of these unusual environmental parameters on embryonic morphometrics may not be direct, but can be mediated through changes in embryonic behavior patterns. For instance, mechanical forces influence formation of bones and cartilage, hence "phenotypic plasticity" of the skeleton (Müller, 2003), so reduced embryonic motility will produce skeletal insufficiencies (Hall and Herring, 1990). The responsiveness of skeletogenesis to embryonic movements means that there is a genetic permissiveness for *de novo* formation of skeletal elements in the embryo, a phenomenon we have often observed while performing experiments in the study of avian craniofacial morphogenesis (Wahl and Noden, 2001).

The variance of maladaptive phenotypic expressions is often greater than "normal". This confounds to some degree the ability to determine whether the response is primary or secondary to the perturbation. One approach utilizes a strain of animals with a mutation in the somatic growth axis as a second control when making morphometric comparisons (Boughner et al., 2008).

3. Morphological responses by embryos and growing vertebrates to environmental variables

Although many environmental variables are known to influence developmental phenotypes, here I will discuss just two: the effects of oxygen tension on overall craniofacial development, and the effects of ambient light on the shape and ultimate size of the developing eye. The reader will find many other examples in the literature on topics such as: the effects of light on pigmentation and neuromast distribution; the effects of environmental organophosphates on limb development, differentiation of the reproductive organs, and rate of sexual maturation; the influence of gravity on early body patterning; and the effects of temperature on sex determination in reptiles.

3.1 Responses to hypoxia

At early embryonic stages, oxygen effects on cell proliferation and differentiation are different from those in the adult. For example, low oxygen tension is critical to certain aspects of normal development and cell differentiation, such as neurulation and chondrogenesis. Each type of embryonic tissue responds uniquely to local variations in oxygen tension (Huang et al., 2004; Webster, 2007). Thus, mesenchymal condensations destined to give rise to endochondral skeletal elements and joints normally show marked hypoxia compared to neighboring tissue during early embryogenesis, and will not differentiate if local O_2 concentrations are too high (Provot et al., 2007; Thompson et al., 1989). Angioblasts, highly migratory cells that aggressively cross tissue boundaries during embryogenesis, retain the ability to switch into a "hypoxic phenotype" as they transition to endothelial cells and adulthood.

Retinopathy of prematurity (ROP), a condition responsible for 13% of the cases of childhood blindness in the U.S. and 62% of the cases in Mexico, occurs because *in utero* blood oxygen levels are much lower than postnatal levels, disturbing vascular development among children born prematurely (Adams, 2008). The effect of this premature "relative hyperoxia" on angiogenesis is to downregulate hypoxia driven, VEGF mediated cell proliferation, resulting in delayed vascularization of the peripheral retina. Subsequent hypoxia in the peripheral retina then produces proliferation of blood vessels in the eye of the premature infant (Fleck and McIntosh, 2008). Children with ROP often display abnormal eye movements and crossed eyes, suggesting that developing periocular tissues are also sensitive to variations in oxygen tension (O'Connor et al., 2007). Also important to this study, a strong correlation has been found between strabismus, anisotropia, amblyopia, and microphthalmia among newborns and maternal smoking during pregnancy (Hakim RB, 1992; Lempert, 2005; Ponsonby AL, 2007; Stone RA, 2006). Smoking lowers maternal blood oxygen carrying capacity because carbon monoxide irreversibly binds to hemoglobin, thus there is a clear implication here that the fetus may be subjected to a relatively hypoxic environment when mom is a smoker.

Thus, hypoxia *per se* can not be said to precipitate abnormal development (Grabowski, 1958), but rather it provokes adaptive changes that occur in response to hypoxia, thereby changing the pattern of gene expression at critical periods (Seta and Millhorn, 2004). Embryonic stem cell populations do not all respond in the same way to hypoxia. The "hypoxic phenotype" among mesenchymal cells is characterized as "highly invasive and expressing several hypoxia regulated genes" (Lash et al., 2002). These features normally characterize trophoblast cells, that are responsive to hypoxic conditions via invasive, migratory behavior—if this behavior fails, abnormal blood flow occurs in the placenta's intervillous spaces as early as week 7 of gestation (Jaffe et al., 1997). Although relatively low oxygen tension is important for proper neurulation, autonomic nerves proliferate excessively along blood vessels among embryos experiencing chronic hypoxia (Ruijtenbeek et al., 2000), thus early peripheral nerve cell populations respond differently to hypoxia from neurepithial cells. Apoptosis and necrosis of brain tissue are among the most dramatic indicators of hypoxia among older embryos (Grabowski, 1966), demonstrating that nervous tissues' response to oxygen changes rapidly as they differentiate and grow. Among myoblasts, the two embryonic processes of cellular division and differentiation show reciprocal behaviors in response to oxygen. Although the rate of *differentiation* of myoblasts as measured by fusion into myotubes is proportional to oxygen concentration, the rate of *division* of myoblasts varies inversely with the oxygen concentration used, within a range of 2%-80% oxygen (Hollenberg et al., 1981).

The behavior of cells during cell migration and differentiation events is critical to proper tissue and organ assembly. A good example of a complex system consisting of different tissues that initially arise and proliferate in isolation from each other, but differentiate and grow in proximity, is the periocular region of the head. Vertebrate vision depends on the ability to stabilize the eye with respect to the surroundings long enough to generate an image on the retina. The oculorotatory muscles that perform this function commit to the myogenic lineage and are hard-wired to the brain very early, before migrating to their final periorbital positions, and while they are still in the paraxial mesoderm along the hindbrain of the embryo (Wahl, 2007). Eye muscles have been observed to develop even in the absence of eyes in some mutants, or where eye size is dramatically reduced (Franz and Besecke, 1991), an indication that the developmental program for early myogenic differentiation is not dependent on the presence of the eye. However, the periorbital environment is where extraocular muscles must integrate with surrounding support tissues and grow to appropriate size, so their ultimate functionality depends on the latter stages of organogenesis. The tissues that support the eyes and share that very limited periocular space include the optic nerve, lacrimal gland, extraocular muscles, fibroadipose tissue, peripheral nerves, ganglionic tissue, and blood vessels. These tissues originate both rostral, caudal, and dorsal to their final location in the periorbital region. They originate as neural crest, neural tube, ectodermal, and mesodermal cells.

3.2 Effects of oxygen deprivation on craniofacial growth in chick embryos

I designed experiments to study the physical environment's effects on early craniofacial development in chick embryos. My preliminary work is described here.

3.2.1 Methods

To learn how acute anoxia affects eye and periocular development, 48 hr chick embryos (Hamburger-Hamilton stages 13-14) were exposed to a pure nitrogen atmosphere at the normal incubation temperature of 38^0 C for 2, 3, or 4 hours as follows: One cc of thin albumen was withdrawn from the pointed end of each egg using a sterile syringe. This eggshell opening was re-sealed with warm paraffin wax. A one-centimeter diameter window over the embryo was made by first cleaning the shell with 70% ethanol, allowing it to dry, and then chipping away the shell using sterile forceps. Embryos were examined and staged according to the Hamburger-Hamilton stage series (HH). Any embryos found to be developing abnormally were eliminated from the experiment, the HH stage of each remaining normal embryo was recorded in pencil on each egg, and the eggshell window was sealed with clear Scotchgard tape. Eggs were transferred to Billups-Rothenburg incubator pods. One pod was flushed for two minutes with high-purity nitrogen gas, and sealed for either 2, 3, or 4 hours. Normal atmospheric air was left in the other (control) pod. Eggs were then returned to a standard, humidified Percival incubator with circulating atmospheric air and allowed to continue developing normally for an additional 2 days. Embryos were examined *in situ*, then collected into 4% paraformaldehyde in phosphate buffer (pH 7.4) for further study.

3.2.2 Results

I found that hypoxia causes craniofacial malformations of increasing severity, proportional to the length of exposure to anoxic conditions (pure nitrogen gas). Most (95%) of both control and experimental embryos survived and were robustly vascularized. A composite photo of representative embryos, placed over a micrometer ruler, is shown in Figure 2. Compared to control embryos (A), 3-hour exposure to anoxic conditions produced ocular phenotypes varying from near-normal to microphthalmic (B), and more than half of all embryos in this group were reduced in size compared to the controls. Four hours of anoxia produced 100% anencephalic, dwarfed embryos (C). Two hours of anoxia resulted in grossly normal embryos (data not shown).

Frontal development of the face in each of these treatments is shown in Figure 3. Normally-developing embryos (A) display prominent medial nasal and maxillary prominences, and the lateral nasal prominence is also well-developed. After 3 hours of anoxia, the maxillary process is reduced or absent (B) and the eyes are smaller than normal. These deformities are not bilaterally symmetrical in every case, as shown in B. The ocular defect includes a lens that is disproportionately large relative to the eyecup. After 4 hours of anoxia, all embryos exhibit anencephalia, but some retain tissues from the lower face (C). At the time of treatment, at stage 14, the primary eye field has already separated, embryos have developed eyecups, and their lens placodes are in the process of invaginating to form vesicles. In C, it can be seen that after 4 hours of anoxia, eyecups subsequently failed to expand. However, lens vesicles did form (arrow). The eyecups differentiated further, but failed to grow: pigmented epithelium extends along the presumptive optic tract. This abnormal distribution of pigmented cells indicates defects in genetic patterning that should have separated the eyecup from the optic tract and brain.

All surviving embryos had well-developed vitelline vasculature, and normal trunk and limb morphology. Several of those exposed to nitrogen had avascular allantoic membranes.

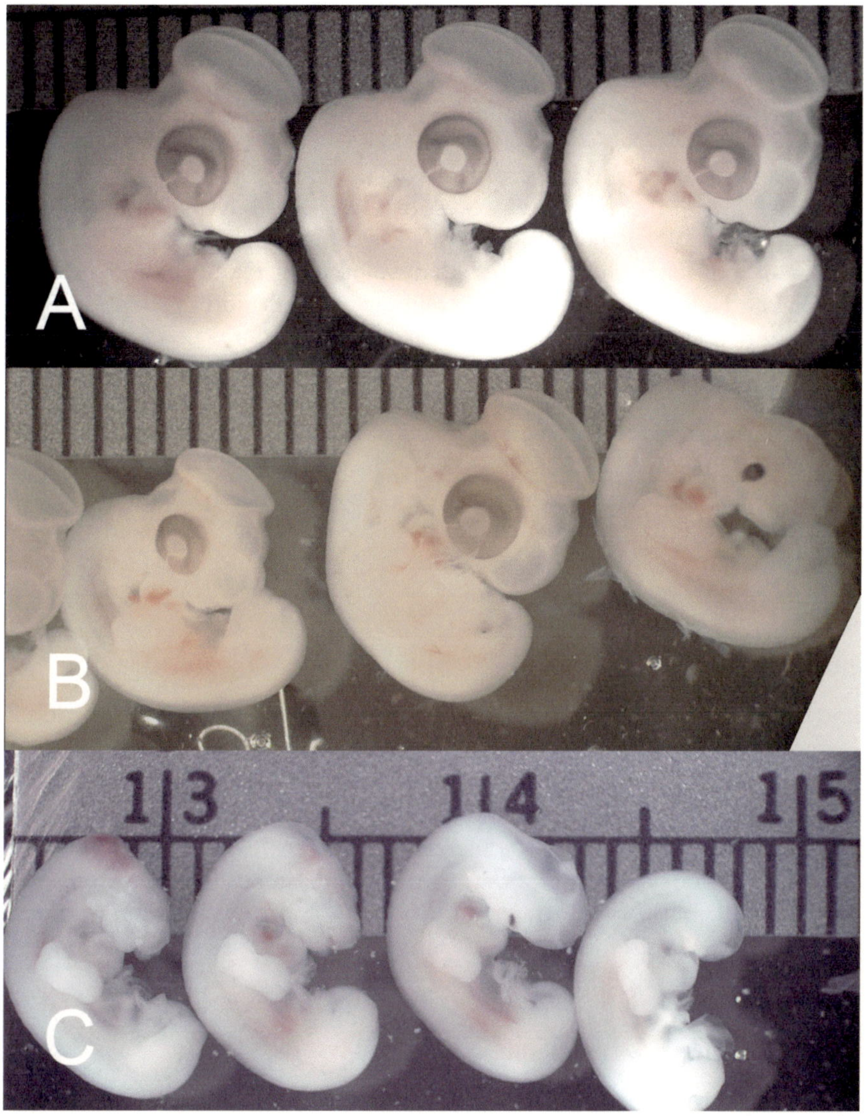

Fig. 2. Lateral view of chick embryos exposed at HH stage 14 (48 hr incubation) to nitrogen gas for 0 (A), 3 hours (B), or 4 hours (C), then returned to normal atmospheric conditions and allowed to develop for a further 2 days. To provide scale, the embryos are positioned over a centimeter ruler.

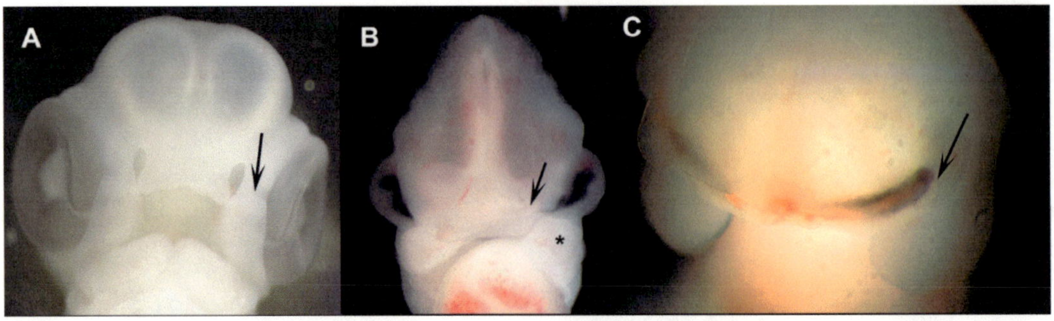

Fig. 3. Frontal views at 4 days: (A) control embryos (HH ~stage 25). Medial and lateral nasal processes are fusing with the maxillary process (arrow). Eyes have expanded greatly since stage 14, when the lens placode was forming a vesicle and inducing formation of the optic cup. (B) Failure of the left maxillary process to develop after 3 hours of anoxia at stage 14. Only the medial and lateral nasal processes are intact (arrow), and the nasal pits are reduced in size. An undivided visceral arch is present (asterisk). Partial fusion of the undivided first arch is seen, with the medial/lateral nasal processes on the right. Reduction in size of the frontonasal prominence of the neural tube, and failure of the eyecups to expand is apparent. Lens vesicles have formed. (C) Following 4 hours of anoxia, the frontonasal prominence is entirely absent. The first arch has not divided (it is located just below the arrow).
The eyecups have failed to expand. Differentiation has proceeded, but genetic programming that should distinguish and separate the eyes from the optic tract has failed, as demonstrated by a trail of pigmented epithelium that extends along the entire presumptive optic tract. Lens vesicles have formed (arrow).

3.2.3 Conclusions

Growth and differentiation of the eyecup, brain, and first visceral arch is retarded if exposed to anoxic conditions at HH stages 13 or 14. The first arch fails to properly divide and grow, resulting in severe reduction of the maxillary process on one or both sides. However, the lens placode does form a lens vesicle and the olfactory placode develops into a nasal pit. Histological assessment is necessary to determine whether this stunted growth is a result of necrosis or arrested mitosis, and to follow the differentiation of the periocular mesenchyme and muscles. Morphometry of the defect at different exposures at different developmental intervals will provide a trajectory of severity that can be analyzed to determine the relative susceptibility of each cell population contributing to the growth of the face.

3.3 Effects of light on growth and shape of the eye

It is a common misperception that the lens and cornea display fixed patterns of development that are independent of non-visual environmental influence, however we have found that light regimen plays an important role in overall shaping of the eye. The effect of light on the cornea, in particular, is of interest because the air/cornea interface is the major focusing surface of the eye. The development of persistent ocular defocus is commonly studied in the chick (Gottlieb et al., 1987; Wallman et al., 1978). Refractive errors (myopia, or nearsightedness, hyperopia, or farsightedness) have been induced in chick eyes using constant darkness (CD, (Gottlieb et al., 1987), and constant light (CL, (Lauber et al., 1970);

(Li, 1995)), producing corneal flattening and hyperopia within three weeks. Long term CL produces shallow anterior chambers, corneal thickening, lenticular thinning, cataracts, and damage to the retina, pigment epithelium, and choroid (Li, 1995).

These studies demonstrate that corneal shaping during growth is influenced by ambient light. We found differences in the pattern of corneal growth between chicks raised in CL vs. normal light conditions (N, raised in 12 hours light/12 hours darkness) using morphometric techniques, including: a) a comparison of eye weights and wet and dry corneal weights, b) measurement of corneal thicknesses and corneal diameters, c) spatial dynamics of corneal expansion, d) measurement of corneal curvatures, and e) stromal cell densities (Wahl, 2009). We learned that the eye's ability to model its shape towards emmetropia is diminished in the absence of periods of light and dark. Particularly sensitive are the stromal cells of the cornea, which show significant changes in density and distribution in CL.

We pursued this finding with additional experiments to learn whether the effect of CL was a direct result of light on the corneal cells, or whether stromal growth of the cornea was regulated by hormones that, in turn, were affected by light cycle (Wahl, 2011). To do this, an organ culture system was designed for chick corneas. Light regimen alone had no effect on corneal growth in culture. Melatonin and/or retinoic acid were applied to the cornea both *in vivo* or *in vitro*, and compared to controls. We found that both melatonin and retinoic acid affect the hydration state of the cornea and alter its shape in growing birds, and we speculate that this effect results from altered ratios of glycosaminoglycans (GAGs) in the corneal matrix. It has been demonstrated that the corneal matrix has a gradient of GAGs that have different properties of hydration (Castoro, 1988), and so it is reasonable to suppose that altering this gradient or changing the ratio of GAG production in any way could affect the curvature of the cornea and its thickness.

4. Morphometric changes in response to physical/mechanical injury

Embryos have a remarkable ability to regenerate themselves through re-specification of cell populations, often resulting in a change in shape and/or body mass. They do not scar, however at birth they are usually smaller and may be physically disproportionate.

4.1 Chick embryo regenerative capacity

Surgical manipulation of avian embryonic tissues always introduces a greater number of variables than the experimenter can control for or, often, readily identify. Because most of our microscopic approaches to the study of embryonic cell behavior, individually or collectively, is limited by the necessity of killing the cells, we really have very little concept about how these cells are dynamically interacting, or what timeframe is involved in those interactions. Most analyses of avian embryonic development are devoted to defining normal events, especially identifying the origins of specific tissues and documenting the precise history and movements of cellular precursors. The observational skills required for this work include morphometric tools that allow interpretation of relationships among tissues surrounding the site or sites of interest. It is important to be prepared for unexpected findings in these studies, as it is all too easy to shoehorn one's observations to fit into a popular theory, rather than consider the possibility that something entirely new is being witnessed.

In the quest to follow the fates of individual precursor cells in chick embryos, the most significant technical advancement was the discovery by Nicole LeDouarin of a nucleolar marker present in most quail cells (Douarin and Barq, 1969). Staining for nucleolar-associated heterochromatin in quail cells allows transplanted quail cells and all their progeny to be followed in avian embryos throughout their development. The quail–chick chimeric method has been applied to nearly all developing organ systems (Wahl and Noden, 2001) but it is in following the fates of highly migratory populations such as the neural crest, myoblasts, angioblasts, and gastrulating mesoblasts that the greatest benefits have accrued.

Morphometric assessment may be made using these methods. Questions such as what number of cellular progeny are produced, how far and in what directions they have moved, and what three-dimensional changes in shape follow a specific time interval or manipulation may be addressed using specific lineage tracing techniques.

One typical method in embryology involves ablation of a target tissue of interest. In many situations, ablations are repaired by compensatory hyperplasia and restitution of the deleted tissue by remaining committed progenitors or adjacent multipotent cells. Healing without restitution, as in the case of an ablated optic cup, may indicate an absence of nearby responsive multipotent populations, or inhibition by newly-differentiated neighboring cells, such as occurs between rhombomeres (Guthrie and Lumsden, 1991). Where restitution takes place, the regenerated element is generally smaller than the normal counterpart, an effect that becomes increasingly pronounced as the age of the embryo at the time of ablation increases.

During transplant procedures, both the size and shapes of the graft and the host lesion sites often change considerably within minutes of excising the tissue. Surface tension at the wound margin contributes to this, both expansive (e.g. surface ectoderm) and compressive (e.g. neural plate). Usually, these changes are transient, however during the initial healing-in time they can be quite important. Many of us have spent hours struggling to fit a curling graft precisely into a well-cut host hindbrain, as the margins of the host site begin to shrink and the graft, too, becomes more compact. If the embryo appears healthy several hours after tissue transplantation and grafted tissue is evident at the intended location, then the surgery is considered a success. If the embryo is alive and shows no gross abnormalities after several days, all the better! However, this 'normalcy' may mask substantial transient or permanent deviations from the normal course of development.

In our experiments on several embryonic tissues in the neurula-stage avian head, we assume that all cells that are directly contacted by microsurgical instruments die immediately. Even among embryos that appear to heal excellently, extensive cellular disintegration adjacent to the lesion is evident via histological examination within a few hours of surgery. This focal cellular trauma can initiate responses that alter the normal intra-embryonic milieu at considerable distances from the site of surgery.

We also found that focal cellular trauma can initiate responses that alter the normal behavior of cells at some distance from the surgical site. In particular, nerve trajectories were disturbed as far away as the forebrain following a lesion in the hindbrain (Wahl and Noden, 2001). Careful morphometry and assessment at multiple times following surgery are important to proper understanding of this phenomenon.

4.2 Salamander regenerative ability following gross physical injury

Most urodeles can regenerate many body tissues, including most of the eye. Structures that regenerate include the retina, the lens, the iris, the pigment epithelium (RPE), and the choroid. Tissue replacement may even be repetitive (Hasegawa, 1965; Reyer, 1977a; Stone, 1960), but the mechanism involved is incompletely understood. Most investigators agree that the central retina is regenerated from the RPE, while the periphery is replaced by cells of the pars-ciliaris-ora serrata complex (Hendrickson, 1964; Keefe, 1973).

During regeneration several processes occur simultaneously, e.g. necrosis triggers phagocytosis by migrating macrophages, the eye's dimensions diminish as the vitreous cavity shrinks, and normally, the lens deteriorates. There is a concurrent proliferation of cell types that ultimately restores function to the eye: cells destined to from new lens, new RPE, and new retina appear and may migrate to sites of continued development. Even cell death among regenerating cells may further affect the changing morphology of the eye (Oppenheim, 1981).

Neural retina regeneration in larval *Triturus. pyrrogaster* and *T. viridescens*, as in adult newts, was initiated primarily at the growth zone of the anterior complex. Larval urodelan eyes, unlike the eyes of adults, are resistant to a temporary loss of blood supply and can be transplanted without a degeneration of the neural retina (Stone, 1930). Regeneration is more rapid in larvae than adults, and is initiated exclusively from the peripheral margins of the retina. In larval *Ambystoma maculatum* lentectomy and retinectomy result in regeneration from the marginal growth zone as in *T. viridescens* (Stone and Cole, 1942; Stone and Ellison, 1945). However, neural retina regeneration did not occur over a waiting period of sixty days when only the retinal pigment epithelium was left in the eye.

Comparing regenerative events of newt limbs with those of the eye is relevant (Zarrow, 1961). During the first stage of limb regeneration, the wound is covered by a specific wound epithelium without which regrowth will not occur. This special epithelium is known as the "wound blastema". A sutured wound will not regenerate...it requires the wound blastema to organize regrowth of the missing tissue(s). Initially, this epithelium is translucent, and later becomes pigmented. It is formed by a single layer of cells that migrate from the periphery of the wound. This regenerative layer later proliferates, becoming up to several cell layers thick, and displays extensive mitotic activity. Initially, the epidermal cells are squamous, becoming columnar as they proliferate. Later they are almost exclusively cuboidal. The basal layers, however, remain low columnar (Zarrow, 1961). The basal cells of the wound epithelium form villous projections into the subjacent dermis. In normal dermis, the reticular basement membrane forms a coarse network through which migratory (macrophage) cells move with relative ease, extending their pseudopodia between the fibers. Epithelization of the wound is followed by a random deposition of fibrils basally that in form resembles a feltlike mat, similar to the normal dermis. Later, lamellar organization and differentiation occur.

5. Morphometry may be the principle way to solve certain developmental problems: 3-D analysis of primordial follicle distribution

Some problems in development are best solved using morphometric analysis. A prime example of this is the ongoing question of ovarian follicular reserves. A "central dogma" of female reproductive biology has long held that oogenesis ceases prior to birth in most mammals and that the functional lifespan of the ovaries is dictated in part by the number of oocytes present; a number that is known to decline precipitously during both fetal development and postnatal life. Primordial (dormant) follicles are distributed in an apparently random fashion throughout the ovarian outer cortex during the three-day estrus cycle of the mouse. However, a discrepancy of 10-fold or more has been shown in the total numbers of follicles among individuals and among mouse strains (Bolon et al., 1997; Bucci et al., 1997). Most consider this a failure of the sampling methods, and call for a more reliable way to evaluate ovarian follicular reserves using a standardized procedure. Most reported methods employ sampling of the ovary by counting representative sections. They use this data to calculate the number of follicles per representative volume, and then multiply that figure by the total volume of the ovary under study(Britt et al., 2004). However, it is difficult to see how one can improve on total sampling of the ovarian reserve, since there are widely different follicle populations among different strains and ages of mice (Myers et al., 2004). This fact, in addition to evidence that replacement germ cells may exist in the bone marrow (Tilly, 2003), suggest that gametes may arise from a more complicated stem cell population than long supposed.

An alternative hypothesis for such variation could be that the population of follicles in the mouse ovary is dynamic, and is in fact replenished by as yet undetermined mechanisms.

We designed a three-dimensional reconstruction method to accurately portray primordial follicle distributions in young mouse ovaries. We reasoned that primordial follicles are not randomly distributed throughout the ovarian cortex, and wished to visualize the variation in follicle distribution in the cortex from ovary to ovary. Our previously unpublished work is presented here.

5.1 Specimen preparation and histological assessment

Three "wild-type mice" were raised until 5 months of age and euthanized by CO_2 overdose during the same stage of estrus. Their left ovaries were fixed in Bouin's fixative and then paraffin embedded. The tissues were serially sectioned at 6μm and stained with Periodic Acid Schiff (PAS) and iron hematoxylin.

Histological examination of the stained and sectioned ovaries showed good preservation of tissue structure and normal ovarian anatomy. However, rare clusters of primordial oocytes sharing a single follicle were found (Figure 4). I have found no reference to the occurrence of follicles with multiple oocytes in the literature. I suspect we found these rare follicles because we were very thorough in our examination of every section from each ovary. We found just two such follicles in the three ovaries reconstructed in Figure 5, and no more than three among several other ovaries not included in this study. These unusual compound follicles are very interesting, although their rarity is an obstacle to further study.

5.2 Analysis of primordial follicular distribution

Using a Zeiss microscope equipped with a camera lucida, tracings of every section within each ovary were made at 100X magnification. Each tracing delineated the boundary of the ovary as well as the location of primordial follicles within that section. Since each follicle occupied more than one section, primordial follicles were defined in this study as those sections containing the nucleus of the oocyte, and surrounded by a single layer of predominantly squamous granulosa cells, of which no more than fifty percent were cuboidal. We used the tracings to map the coordinates of each primordial follicle in three-dimensional space using SYSTAT with the ovary slice number as the Z coordinate.

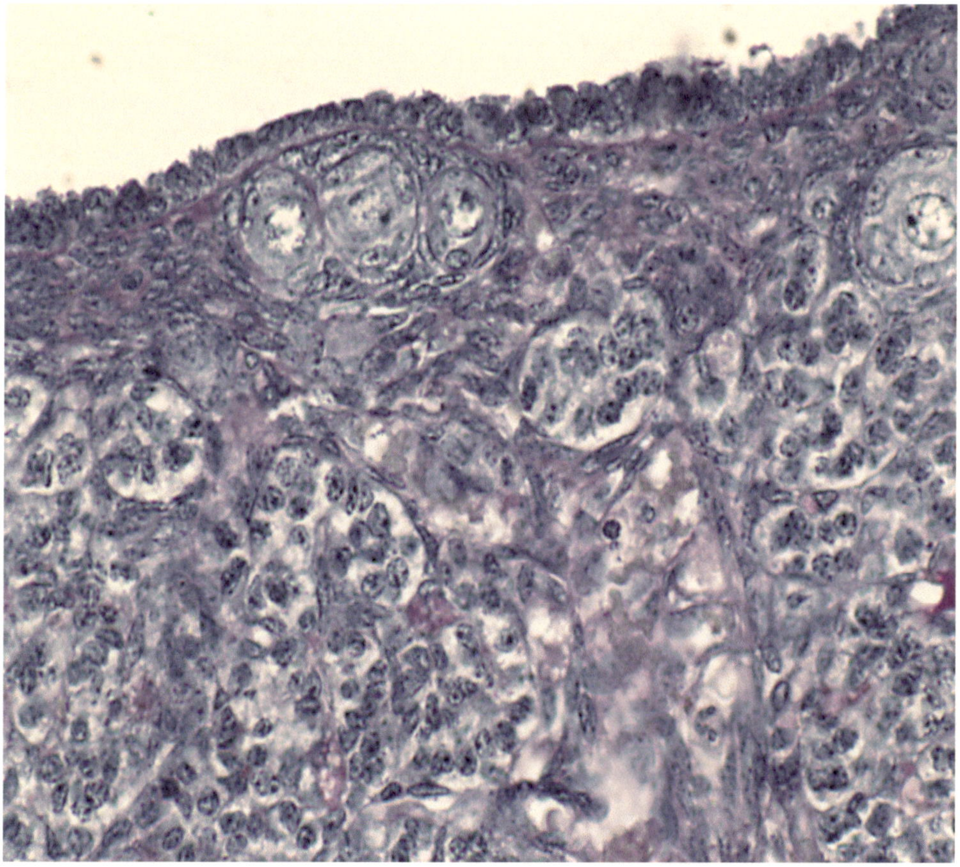

Fig. 4. Cluster of three oocytes within a single primordial follicle from a 6 micron paraffin section stained with iron hematoxylin and PAS. Primordial oocyte clusters are not discussed in the literature, however we see them occasionally. They are usually located near the germinal epithelium, as seen here.

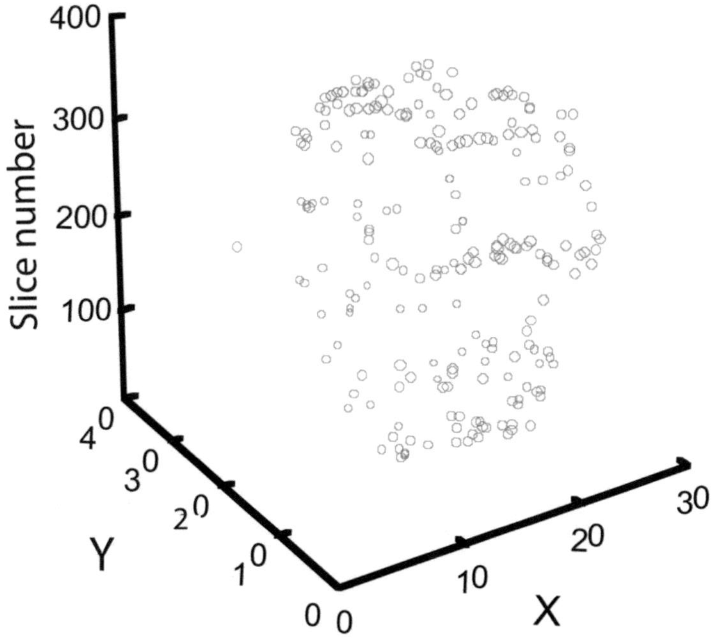
400
300
200
100
Slice number
40
30
20
10
Y
0
0
10
20
30
X

A

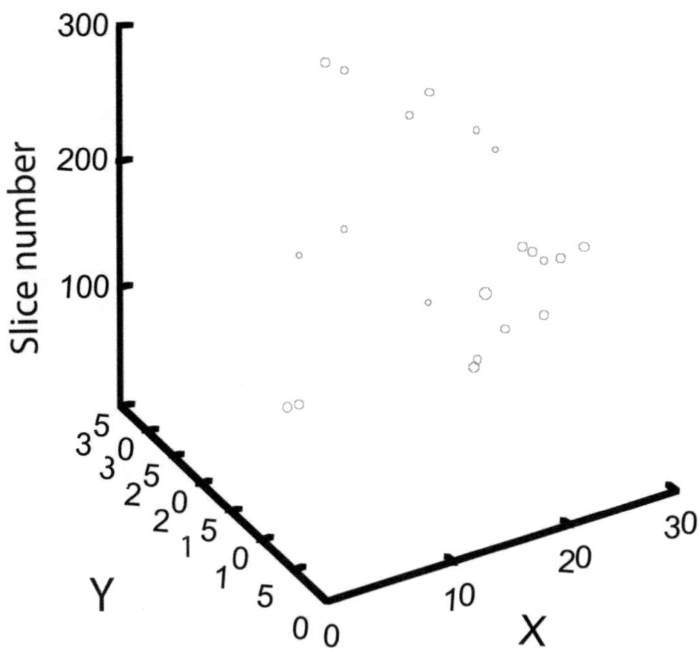
300
200
100
Slice number
35
30
25
20
15
10
5
Y
0
0
10
20
30
X

B

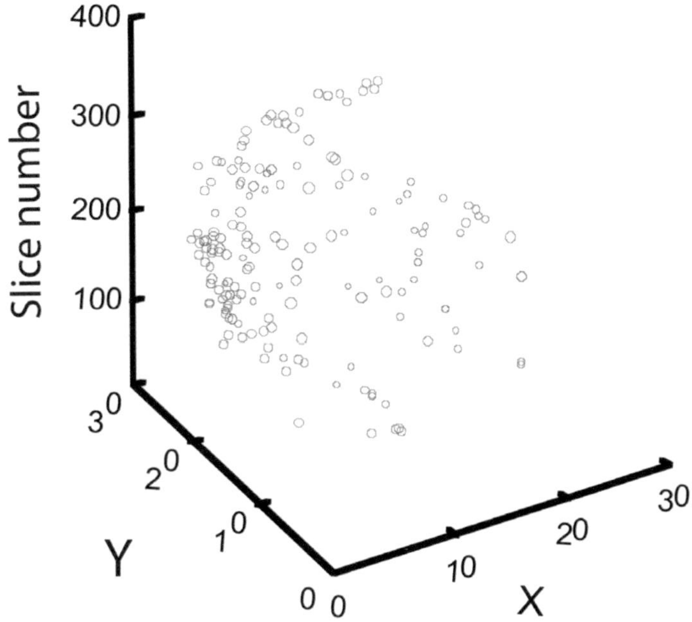

C

Fig. 5. 3-D reconstruction of primordial follicle distribution within three mouse ovaries. Ovary A. Coefficient of Dispersion = 2.30, Ovary B. Coefficient of Dispersion = 1.28, Ovary C. Coefficient of Dispersion =4.05. In this representation, follicles that are near the observer in the Z axis are shown as large circles, whereas those further away are small circles. Note that ovary B has far fewer follicles than either A or C, although all three were from 5 month old female mice that came from the same litter. Note that none of these have random distributions of follicles, but rather, the follicles occur in clumped patterns.

Each ovary drawing was divided into approximate cubic units (350 X 350 X 300 μm), and then the number of primordial follicles in each cube was recorded, discounting cubes containing the medulla or corpora luteae. We graphed the positions of primordial follicles among cubes using a modified Poisson distribution, and then calculated a coefficient of dispersion by finding the ratio of the variance of numbers of follicles per cube to the mean number per cube.

A coefficient of dispersion greater than 1 is indicative of a clumped distribution pattern. All three ovaries had a coefficient of dispersion greater than 1, thus we conclude that primordial follicles are non-randomly distributed in the ovarian cortex. Also apparent from the figures is that the number of primordial follicles varies widely from one ovary to the next.

From just these three reconstructions, it may be seen that a) there are not enough primordial follicles in the 5 month mouse ovary to account for the number required throughout its reproductive lifespan, b) primordial follicles are not randomly dispersed throughout the ovary, and c) numbers of follicles vary widely from one mouse to the next, although all three were collected while in the same phase of the estrus cycle. These observations demonstrate the power of careful morphometry in elucidating important, fundamental facts about the basic biology of the organism that are difficult to obtain any other way.

6. Summary

In this paper, I have tried to provide a sense of the great range of morphological plasticity in developing systems…both plasticity of normal development, and in response to injury or environmental change. In addition to a brief review of the literature, I have used examples from my own work (both published and unpublished) to illustrate the plasticity of form among a wide variety of vertebrate embryos, and I have indicated how morphometric analysis is a useful tool for learning about the changes of form possible in developing vertebrates. The emergent properties of the developing organism, both in response to the environment, or following injury, illustrate yet again that in biological systems, the whole is always greater than the sum of its parts.

7. Acknowledgments

I wish to thank the following individuals who participated in different parts of the original research projects reported here: Howard C. Howland, Dept. of Neurobiology and Behavior, Cornell University, Allison Inga, former Wells College undergraduate and now student of veterinary medicine at Ross University, Drew M. Noden, Department of Biomedical Sciences, Cornell University, and Yuko Takagi, former Wells College undergraduate and now postdoctoral associate at Harvard University School of Medicine. I also thank Wells College for providing space and opportunities for undergraduates to pursue original research, and the NIH for funding that supported my work on eye growth. Finally, I wholeheartedly thank my husband, Ellis Loew, Department of Biomedical Sciences, Cornell University, for help in reviewing this chapter, and for tolerating household mayhem while I wrote it.

8. References

Adams G (2008) Editorial on retinopathy of prematurity. Early Human Development 84:75-76

Bolon B, Bucci TJ, Warbritton AR, Chen JJ, Mattison DR, Heindel JJ (1997) Differential follicle counts as a screen for chemically induced ovarian toxicity in mice: results from continuous breeding bioassays. Fundamentals of Applied Toxicology 39:1-10

Bonner JT (1969) On Growth and Form, by D'Arcy Wentworth Thompson. Cambridge University Press

Boughner JD, Wat S, Diewert VM, Young NM, Browder LW, Hallgrímsson BJ (2008) Short-faced mice and developmental interactions between the brain and the face. Journal of Anatomy 213:646-662

Britt KL, Ebling FJP, Kerr JB, Myers M, Wreford NGM (2004) Methods for quantifying follicular numbers within the mouse ovary. Reproduction 127:569-580

Bucci TJ, Bolon B, Warbritton AR, Chen JJ, Heindel JJ (1997) Influence of sampling on the reproducibility of ovarian follicle counts in mouse toxicity studies. Reproductive Toxicology 11:689-696

Buresova M, Zidek V, Musilova A, Simakova M, Fucikova A, Bila V, Kren V, Kazdova L, Di Nicolantonio R, Pravenec M (2006) Genetic relationship between placental and fetal weights and markers of the metabolic syndrome in rat recombinant inbred strains Physiol Genomics 26:226-231

Castoro JA, A. A. Bettelheim, et al (1988) Water gradients across bovine cornea. Investigative Opthalmology and Visual Science 29:963

Douarin NML, Barq G (1969) Use of Japanese quail cells as 'biological markers' in experimental embryology. C R Acad Sci Hebd Seances Acad Sci D 269:1543-1546

Etcoff N (2000) Survival of the Prettiest: The Science of Beauty. Abacus Books, London

Fleck B, McIntosh N (2008) Pathogenesis of retinopathy of prematurity and possible preventive strategies. Early Human Development 84:83-88

Franz T, Besecke A (1991) The development of the eye in homozygotes of the mouse mutant extra- toes. Anat EMBRYOL 184:355-362

Gottlieb M, Wentzek L, Wallman J (1987) Different visual restrictions produce different ametropia and different eye shapes. Investigative Ophthalmology and Visual Science 28:1225-1235

Gould SJ (1966) Allometry and size in ontogeny and phylogeny. Biological Reviews 41:587-640

Grabowski CT (1966) The etiology of hypoxia-induced malformations in the chick embryo. Journal of Experimental Zoology 157:307-326

Grabowski CT, Paar, John A. (1958) The teratogenic effects of graded doses of hypoxia on the chick embryo. The American Journal of Anatomy 103:313-347

Guthrie S, Lumsden A (1991) Formation and regeneration of rhombomere boundaries in the developing chick hindbrain. Development 112:221-229

Hakim RB TJ (1992) Maternal cigarette smoking during pregnancy. A risk factor for childhood strabismus. Arch Ophthalmol 110:1459-1462

Hall BK, Herring SW (1990) Paralysis and growth of the musculoskeletal system in the embryonic chick. Journal of Morphology 206:45-56

Hasegawa M (1965) Restitution of the eye from the iris after removal of the retina and lens together with the eye-coats in the newt, Triturus pyrrhogaster. Embryologia 8:362-386

Hendrickson A (1964) Regeneration of the retina in the newt Diemictylus v. viridescens. University of Washington

Hollenberg M, Honbo N, Ghani QP, Samorodin AJ (1981) Oxygen enhances fusion of cultured chick embryo myoblasts. . Journal of Cellular Physiology 106:1097-4652

Huang S-TJ, Vo KCT, Lyell DJ, Faesen GH, Tulac S, Tibshirani R, Giaccia AJ, Giudice LC (2004) Developmental response to hypoxia. The FASEB Journal 18:1348-1365

Jaffe R, Dorgan A, Abramowicz JS (1997) Maternal circulation in the first-trimester human placenta--myth or reality? American Journal of Obstetrics and Gynecology 176:695-705

Keefe JR (1973) An analysis of urodelian retinal regeneration: IV. Studies of the cellular source of retinal regeneration in Triturus cristatus carnifex using H3-thymidine. Journal of Experimental Zoology 184:239-258

Klingenberg CP (2003) A developmental perspective on developmental instability: theory, models, and mechanisms. In: Polak M (ed) Developmental Instability: Causes and consequences Oxford Press, New York

Klingenberg CP (2008) Morphological integration and developmental modularity. Annu Rev Ecol Evol Syst 39:115-132

Lash GE, Postovit L-M, Matthews NE, Chung EY, Canning MT, Pross H, Adams MA, Graham CH (2002) Oxygen as a regulator of cellular phenotypes in pregnancy and cancer. Canadian Journal of Physiology and Pharmacology 80:103-109

Lauber J, Boyd J, Boyd T (1970) Intraocular pressure and aqueous outflow facility in light-induced avian buphthalmos. Experimental Eye Research 9:181-187

Lempert P (2005) Amblyopia pervalence and cigarette smoking by women. Opthalmic Physiol Opt 25:592-595

Li T, D. Troilo, et al (1995) Constant light produces severe corneal flattening and hyperopia in chickens. Vision Research 35(9):1203-1209

McCune AR (1981) Quantitative description of body form in fishes: Implications for species level taxonomy and ecological inference. Copeia 1981:897-901

Mitashov VI (1978) Replacement of melanin granules in the iris and pigment epithelium of the retina in adult newt after completion of eye regeneration. Soviet Journal of Developmental Biology 9:150-155

Müller GB (2003) Embryonic motility: environmental influences and evolutionary innovation. Evolution and Development 5:56-60

Müller GB (2011) BIO. Evolution and Development 13:243-246

Myers M, Britt KL, Wreford NGM, Ebling FJP, Kerr JB (2004) Methods for quantifying follicular numbers within the mouse ovary. Reproduction 127:569-580

O'Connor AR, Wilson CM, Fielder AR (2007) Opthalmological problems associated with preterm birth. Eye 21:1254-1260

Oppenheim RW (1981) Neuronal cell death and some related regressive phenomena during neurogenesis: A selective historical review and progress report. In: Cowan WM (ed) Studies in developmental neurobiology; essays in honor of Viktor Hamburger. Oxford Press, N.Y., pp 74-133

Ponsonby AL BS, Kearns LS, MacKinnon JR, Scotter LW, Cochrane JA, Mackey DA (2007) The association between maternal smoking in pregnancy, other early life characteristics and childhood vision: the Twins Eye Study in Tasmania. Ophthalmic Epidemiol 14:351-359

Provot S, Zinyk D, Gunes Y, Kathri R, Le Q, Kronenberg HM, Johnson RS, Longaker MT, Giaccia AJ, Schipani E (2007) Hif-1a regulates differentiation of limb bud mesenchyme and joint development. Journal of Cell Biology 177:451-464

Reyer R (1977a) Repolarization of reversed, regenerating lenses in adult newts Notophthalmus viridescens. Experimental Eye Research 24:501-509

Reyer RW (1977b) The amphibian eye: Development and regeneration. Handbook of Sensory Physiology, pp 309-390

Ruijtenbeek K, le Noble FAC, Janssen GMJ, Kessels CGA, Fazzi GE, C.E. B, De Mey JGR (2000) Chronic hypoxia stimulates periarterial sympathetic nerve development in chicken embryo. Circulation Research 102

Serrat MA, King D, Lovejoy CO (2008) Temperature regulates limb length in homeotherms by directly modulating cartilage growth. PNAS 105:19348-19353

Seta KA, Millhorn DE (2004) Functional genomics approach to hypoxia signaling. Journal of Applied Physiology 96:765-773

Stone LS (1930) Heteroplastic transplantation of eyes between the larvae of two species of Amblystoma. J Exp Zool 55:193-261

Stone LS (1960) Regeneration in the lens, iris, and neural retina in a vertebrate eye. Yale Journal of Biology and Medicine 32:464-473

Stone LS, Cole CH (1942) Grafted eyes of young and old adult salamanders (Amblystoma punctatum) showing return of vision. Yale Journal of Biolgy and Medicine 15:735-754

Stone LS, Ellison FS (1945) Return of vision in eyes exchanged between adult salamanders of different species. Journal of Experimental Zoology 100:217-227

Stone RA WL, Ying GS, Liu C, Criss JS, Orlow J, Lindstrom JM, Quinn GE (2006) Associations between childhood refraction and parental smoking. Invest Ophthalmol Vis Sci 47:4277-4287

Thompson TJ, Owens PDA, Wilson DJ (1989) Intramembranous osteogenesis and angiogenesis in the chick embryo. Journal of Anatomy 166:55-65

Tilly JL (2003) Ovarian follicle counts--not as simple as 1,2,3. Reproductive Biology and Endocrinology 1:11

Van Valen L (1962) A study of fluctuating asymmetry. Evolution 16:125-142

Wahl C (2007) Periocular Mesenchyme: The interactions of neural crest and mesoderm. In: Tasman W, Jaeger EA (eds) Duane's Foundations of Clinical Ophthalmology. Harper and Row

Wahl CM (1985) SEM study of photoreceptor differentiation in regenerating retinas of the newt N. viridescens. Physiology. Cornell, Ithaca, NY

Wahl CM, Li, T.,Choden,T., Howland, H.C. (2009) Morphometrics of corneal growth of chicks raised in constant light. Journal of Anatomy 214:355-361

Wahl CM, Noden DM (2001) Cryptic responses to tissue manipulations in avian embryos. International Journal of Developmental Neuroscience 19:183-196

Wahl CM, T. Li, et al (2011) Effects of light and melatonin on chick corneas grown in culture. Journal of Anatomy in press

Wallman J, Turkel J, Trachtman J (1978) Extreme Myopia Produced by Modest Change in Early Visual Experience. Science 201:1249-1251

Webster KA (2007) Hypoxia: Life on the Edge. Antioxidants & Redox Signaling 9:1303-1307

Wells JC (2007) The thrifty phenotype as an adaptive maternal effect. Biol Rev Camb Philos Soc 82:143-172

Yamada T, Dumont JN, Moret, Brun (1978) Autophagy in dedifferentiating newt iris epithelial cells in vitro. Differentiation 11:113-147

Zarrow MXe (1961) Growth in Living Systems. Basic Books, New York.